Passing the Numeracy Skills Test

Revised fifth edition

Mark Patmore

Los Angeles | London | New Delhi
Singapore | Washington DC

Learning Matters
An imprint of SAGE Publications Ltd
1 Oliver's Yard
55 City Road
London EC1Y 1SP

SAGE Publications Inc.
2455 Teller Road
Thousand Oaks, California 91320

SAGE Publications India Pvt Ltd
B 1/I 1 Mohan Cooperative Industrial Area
Mathura Road
New Delhi 110 044

SAGE Publications Asia-Pacific Pte Ltd
3 Church Street
#10-04 Samsung Hub
Singapore 049483

Editor: Amy Thornton
Production controller: Chris Marke
Project management: Deer Park Productions
Marketing manager: Catherine Slinn
Cover design: Topics – The Creative Partnerhsip
Typeset by: PDQ Typesetting Ltd
Printed and bound in Great Britain by: MPG Books
Group, Bodmin, Cornwall

Library of Congress Control Number: 2012935898

British Library Cataloguing in Publication data

A catalogue record for this book is available from the
British Library.

ISBN 978 1 4462 7515 3 (hbk)

ISBN 978 1 4462 7516 0 (pbk)

Passing the Numeracy Skills Test

Revised fifth edition

Contents

Acknowledgements

The glossary is reproduced courtesy of the TA © Teaching Agency for Schools. Permission to reproduce TA copyright material does not extend to any material which is indentified as being the copyright of a third party or any photographs. Authorisation to reproduce such material would need to be obtained from the copyright holders.

About the author

Mark Patmore is a former senior lecturer in mathematical education at the School of Education at Nottingham Trent University. He is currently working in teacher education with other universities and training providers and also provides CPD for teachers of mathematics. After some years as a numeracy consultant for the Teacher Training Agency, Mark was for several years one of the writers for the numeracy skills tests. Mark is now a member of the Test Review Group managed by Alpha*Plus* Consultancy which monitors the writing of the tests.

Mark is a principal assessor for the Functional Skills Tests for a leading examination board and is involved with assessing and verifying a range of educational qualifications.

He is the author or co-author of a number of publications for GCSE and Key Stage 3.

Series introduction

The QTS Skills tests

All new entrants to the teaching profession in England, including those on initial teacher training (ITT) and Graduate Teacher programmes (GTP), have to pass the skills tests to be eligible for the award of qualified teacher status (QTS).

Following the Department for Education (DfE) publication, *Training our next generation of outstanding teachers: Implementation plan* (November 2011), the requirements for trainees starting courses in the 2012 academic year remain unchanged; candidates are required to pass the skills tests in their final year of study (with the current exceptions for those on flexible routes into teaching).

The tests cover skills in:

- **numeracy, and**
- **literacy.**

Trainees starting a course from September 2013 will be required to pass pre-entry skills tests in numeracy and literacy before starting their course. The pass mark for these tests will be raised and the number of resits allowed will be limited to two.

The tests will demonstrate that you can apply these skills to the degree necessary for their use in your day-to-day work in a school, rather than the subject knowledge required for teaching. The tests are taken online by booking a time at a specified centre, are marked instantly and your result, along with feedback on that result, will be given to you before you leave the centre.

You will find more information about the skills tests and the specified centres on the Teaching Agency (TA) website: *www.education.gov.uk*

Titles in this series

This series of books is designed to help you become familiar with the skills you will need to pass the tests and to practise questions on each of the topic areas to be tested.

Passing the Numeracy Skills Test (revised fifth edition)
Mark Patmore
ISBN 978 1 4462 7515 3 (hbk)
ISBN 978 1 4462 7516 0 (pbk)

Passing the Literacy Skills Test (third edition)
Jim Johnson and Bruce Bond
ISBN 978 0 85725 879 3
ISBN 0 978 1 44625 682 4 (hbk)

Introduction

Introduction to the test

The numeracy skills test is a computerised test, which is divided into two sections:

- **section 1 for the mental arithmetic questions;**
- **section 2 for the written questions known as 'on-screen' questions.**

The **mental arithmetic** section is an audio test heard through headphones. Calculators are not allowed but noting numbers and jotting down working will be permitted. There are 12 questions in this section. Note: this section must be answered first; each question has a fixed time in which you must answer; you cannot return to a question if you later wish to change your answer. Questions will be asked to test your ability to carry out mental calculations using fractions, percentages, measurement, conversions and time.

The **'on-screen'** questions: there are 16 written questions in this section and they could be asked in any of the following formats.

- **Multiple choice questions where you will choose the correct answer from a number of possible answers.**
- **Multiple response questions where you will have to choose more than one correct statement from a number of possible statements.**
- **Questions requiring a single answer.**
- **Questions where you will have to 'point and click' on the correct response. (To change an answer click on an alternative area of the table or graph.)**
- **Questions where you will select and place your chosen answer into selected boxes. (To change an answer drag it back to its original position and choose another.)**

In this part of the test you can use the 'on-screen' calculator. You answer questions using the mouse and the keyboard and you can move between questions by using the 'next' and 'previous' buttons. You can return to questions either by using the 'flag' button and then the 'review' button or by waiting until the end of the test when you will have the option of reviewing all the questions – provided there is time!

The 'on-screen' part of the test will ask questions covering the following two areas.

- **Interpreting and using statistical information.**
  These questions will test your ability to identify trends correctly, to make comparisons and draw conclusions and interpret charts and tables correctly.

There are 7 questions in this area.

- **Using and applying general arithmetic.**
  The questions in this area will test your ability to use general everyday arithmetic correctly using, e.g. time, money, ratio and proportion, fractions, decimals and percentages, distance and area, conversions between currencies, simple formulae and averages, including mean, median, mode and range.

There are 9 questions in this area.

Time for the test

The time limit for the whole test is approximately 48 minutes at the end of which the test will shut down automatically.

The contexts for the questions

One of the aims of the numeracy skills test is to ensure that teachers have the skills and understanding necessary to analyse the sort of data that is now in schools and consequently most questions will be set within contexts such as:

- **national test data;**
- **target setting and school improvement data;**
- **pupil progress and attainment over time;**
- **special educational needs;**
- **GCSE subject choices and results.**

Hints and advice

The mental arithmetic, audio, section

Each mental arithmetic question is heard twice. After the first reading an answer box appears on the screen. You will have a short time to type an answer into the box, after which the next question will automatically appear. As mentioned earlier, you cannot move forwards or backwards between questions. At the start of the test a practice question will be heard.

- **Concentrate the first time the question is read, and note down the key numbers. For example, a question could be 'In a class of 30 pupils 24 are boys, what fraction are girls?'. You should jot down 30 and 24. The second time the question is heard concentrate on what to do with those numbers (e.g. $30 - 24 = 6$; $6/30 = 1/5$).**

- **Start to work out the answer as soon as you have the information. If possible, you may be able to do this while you hear the second reading of the question.**

- **If you cannot answer a question don't worry or panic – enter a likely answer then forget it. Remember, you don't need to get every single question right.**

- **Note that you don't have to worry about units, i.e. £ or cm. The units will appear in the answer box.**

- **Listen carefully to what the question requires in the answer. For example, a question could ask for a time 'using the 24-hour clock', or an answer 'to the nearest whole number', or 'to two decimal places'. (There are notes on this in Chapter 3.)**

- **Fractions need to be entered in the lowest terms. For example, 6/8 should be entered as 3/4 and 7/28 should be entered as 1/4.**

- **Practise using mental strategies. For example, purchasing five books that cost £5.99 can be worked out by multiplying 5 × £6 (£30) and subtracting 5 × 1p (5p) to give £29.95.**

- **Remember the link between fractions and percentages – see Chapter 1.**

The on-screen questions

- Try not to spend more than two minutes on any one question and keep an eye on the time remaining. If you think you are exceeding the time then move on – you can always return to any you still need to complete at the end of the test and insert an answer. Try not to leave any answers blank at the end of the test.

- Read each question carefully. For example, a question may ask for the percentage of pupils who achieved level 4 and above. Don't just look at those who gained level 4; you need to include those who achieved level 5, level 6, and so on.

- Check that you are giving the correct information in the answer. A table may give you details of the number of marks a pupil achieved but the question may be asking for a percentage score.

The on-screen calculator

When the mental arithmetic section of the test is finished, a basic 4-function calculator will be available on the screen for you to use for the rest of the test. No other calculators are to be used. You can move the calculator around the screen using the mouse.

The on-screen calculator works through the mouse and through the number pad on the keyboard. If you wish to use the number pad you must ensure that the number lock key 'Num Lock' is activated.

Notes on using the on-screen calculator

- To cancel an operation, press CE .

- Always use the 'clear' button C on the calculator before beginning a new calculation.

- Always check the display of the calculator to make sure that the number shown is what you wanted.

- Check calculations and check that your final answer makes sense in the context of the question. For example, the number of pupils gaining level 5 in a test will not be greater than the size of the cohort or group.

Other hints

1. Rounding up and down

- Make sure that any instructions to round an answer up or down are followed or the answer will be marked as incorrect.

- Use the context to make sure whether a decimal answer should be rounded up or down. For example, an answer of 16.4 lessons for a particular activity is clearly not appropriate and the answer would need to be rounded up to 17 lessons.

- Questions may specify that the answer should be rounded to the nearest whole number or be rounded to two decimal places. See the notes at the start of Chapter 3.

- When carrying out calculations relating to money, the answer shown on the calculator display will need to be rounded to the nearest penny (unless otherwise indicated). Hence, if calculating in pounds, round to two decimal places to show the number of pence. If 10.173 is the answer in pounds on the calculator display, rounding to the nearest penny gives £10.17.

2. Answering multi-stage questions

The calculator provided is not a scientific calculator and therefore care needs to be taken with 'mixed operations', i.e. calculations using several function keys. It is important that the function keys are pressed in the appropriate order for the calculation. It may also be useful to note down answers to particular stages of the calculation.

It is important to remember to carry out the calculation required by the question in the following order: any calculation within brackets followed by division/multiplication followed by addition and/or subtraction. Thus, the answer to the calculation $2 + 3 \times 4$ is 14 and not 20, the answer to $\frac{18}{3+6}$ is $\frac{18}{9} = 2$ and the answer to $\frac{18}{3} + 6 = 6 + 6 = 12$. See the notes at the start of Chapter 3.

3. Dealing with fractions

To enter fractions, use the division sign, e.g. for $\frac{5}{8}$ of 350, enter $5 \div 8$ to reach 0.625, then multiply by 350 to reach 218.75.

How to use this book

The book is divided into 6 chapters:

Chapter 1: a very short chapter included to remind you of the basic arithmetic processes. The majority of you will be able to miss this unit out, but some may welcome a chance to revise fractions, decimals and percentages, etc.

Chapters 2–4: these cover the three 'content' areas (see above), one area per chapter.

Chapter 5: a practice mental arithmetic test, and a full practice onscreen test for you to work through.

Chapter 6: answers and key points for all the questions in the main chapters, and for the sample tests.

In each chapter, the additional required knowledge, language and vocabulary are explained, and worked examples of the type of questions to be faced are provided together with the practice questions. The answers for these questions are given at the end of the book, together with further advice and guidance on solutions.

Revision checklists

The following charts show in detail the coverage of the three main chapters and the practice tests. You can use the checklists in your revision, to make sure that you have covered all the key content areas.

Revision checklist for Chapter 2: Mental arithmetic

Syllabus Reference	Content	Question
1a	Time – varied contexts	1, 7, 10, 18, 20, 21, 34, 36, 39, 40
1b	Amounts of money varied contexts	12, 38
1c/d/e	1c Proportion – answer as a fraction 1d Proportion – answer as a percentage 1e Proportion – answer as a decimal	
1f	Fractions	16, 29
1g	Decimals	32
1h	Percentages – varied contexts	2, 4, 9, 11, 13, 19, 22, 24, 25, 26, 27, 28, 30, 33, 35, 37
1i/j/k	1i Measurements – distance 1j Measurements – area 1k Measurements – other	23
1l/m/n/o/p/q	1l Conversions from one currency to another 1m Conversions – from fractions to decimals 1n Conversions – from decimals to fractions 1o conversions – from percentages to fractions 1p Conversions – from fractions to percentages 1q Conversions – other	15 31 6
1r	Combination of one or more of addition, subtraction, multiplication, division (may involve amounts of money or whole numbers)	3, 5, 8, 14, 17

Revision checklist for Chapter 3: Using and applying general arithmetic

(Note: Only the main references are used; many questions will cover more than 1 reference)

Syllabus Reference	Content	Question
3a	Time – varied contexts	7, 17, 23, 29, 31, 33, 37
3b	Amounts of money	2
3c, d, e, f	3c Proportion – answer as a fraction 3d Proportion – answer as a percentage 3e Proportion – answer as a decimal 3f Ratios	14 14 14 3, 59
3g	Percentages – varied contexts	4, 5, 6, 8, 9, 10, 11, 12, 13, 18, 27, 39, 41, 42, 47, 48, 60
3h	Fractions	19, 24, 54
3i	Decimals	
3j, k, l, m, n	3j Measurements – distance 3k Measurements – area 3l Conversions – from one currency to another 3m Conversions – from fractions to decimals or vice versa 3n Conversions – other	21, 26, 28, 30, 35, 36 22 15, 16, 20, 24, 34, 38
3o, p, q, r, s	3o Averages – mean 3p Averages – median 3q Averages – mode 3r Range 3s Averages – combination	45, 58 1, 44
3t	Given formulae	32, 40, 43, 46, 49, 50, 51, 52, 53, 55, 56, 57

Revision checklist for Chapter 4: Interpreting and using statistical information

(Note: Only the main references are used; many questions will cover more than 1 reference)

Syllabus Reference	Content	Question
2a	Identify trends over time	13, 17, 19
2b	Make comparisons in order to draw conclusions	2, 5, 10, 11, 12, 14
2c	Interpret and use information	1, 3, 4, 6, 7, 8, , 15, 16, 18, 20
3o	Averages	9

Revision checklist for the practice mental arithmetic test

Syllabus Reference	Content	Question
1a	Time – varied contexts	6
1b	Amounts of money – varied contexts	7, 10
1c/d/e	1c Proportion – answer as a fraction 1d Proportion – answer as a percentage 1e Proportion – answer as a decimal	
1f	Fractions	12
1g	Decimals	
1h	Percentages – varied contexts	2, 3, 5, 8
1i/j/k	1i Measurements – distance 1j Measurements – area 1k Measurements – other	
1l/m/n/o/p/q	1l Conversions – from one currency to another 1m Conversions – from fractions to decimals 1n Conversions – from decimals to fractions 1o conversions – from percentages to fractions 1p Conversions – from fractions to percentages 1q Conversions – other	9 4
1r	Combination of one or more of addition, subtraction, multiplication, division (may involve amounts of money or whole numbers)	1, 11

Revision checklist for the practice on-screen test

Syllabus Reference	Content	Question
2a	Identify trends over time	4, 9
2b	Make comparisons in order to draw conclusions	1, 8
2c	Interpret and use information	2, 6, 7
3a	Time – varied contexts	
3b	Amounts of money	3, 10, 13
3c, d, e, f	3c Proportion – answer as a fraction 3d Proportion – answer as a percentage 3e Proportion – answer as a decimal 3f Ratios	5, 14
3g	Percentages – varied contexts	15, 16
3h	Fractions	
3i	Decimals	
3j, k, l, m, n	3j Measurements – distance 3k Measurements – area 3l Conversions – from one currency to another 3m Conversions – from fractions to decimals or vice versa 3n Conversions – other	
3o, p, q, r, s	3o Averages – mean 3p Averages – median 3q Averages – mode 3r Range 3s Averages – combination	
3t	Given formulae	9, 11

1 | Key knowledge

Fractions, decimals and percentages

You must remember decimals and place value:

hundreds	tens	ones	•	tenths	hundredths
4	3	5	•	2	7

4 hundreds + 3 tens + 5 ones + 2 tenths + 7 hundredths

$$= 400 + 30 + 5 + \frac{2}{10} + \frac{7}{100}$$

$$= 435.27$$

Take care when adding or subtracting decimals to line up the decimal points. Remember, too, when multiplying decimals by 10 that all the digits move one place to the left, so 435.27 × 10 becomes 4352.7, and when dividing by 100 the digits move two places to the right so 4352.7 ÷ 100 becomes 43.527.

When multiplying two decimals the method you may remember is to 'ignore' the decimal points, do the multiplication and then count up the number of decimal figures in the question numbers – the total will give the number of decimal figures in the answer number, so that 0.4 × 0.5 is calculated as 4 × 5 = 20; there are 2 decimal figures in the question numbers (4 and 5) so there are 2 in the answer. Therefore the answer is 0.20. It would be better to think of this calculation as follows:

$$0.4 \times 0.5 = \frac{4}{10} \times \frac{5}{10} = \frac{20}{100} = 0.2$$

You must remember how to work with fractions. There are several ways of 'looking' at a fraction, for example: $\frac{3}{4}$ = 3 parts out of 4; or 3 divided by 4 or 3 shared by 4 = 3 ÷ 4 = 0.75; or three lots of a quarter = $3 \times \frac{1}{4}$; or a quarter of 3 = $\frac{1}{4} \times 3$.

One way to calculate, say, $\frac{2}{5}$ of £20 is: find a fifth, £20 ÷ 5 = £4 then multiply this by 2 = £8. Another way is to change the fraction into a decimal:

$\frac{2}{5}$ = 2 ÷ 5 = 0.4, then multiply 0.4 × £20 = £8.

You will need to know how to simplify a fraction by dividing both the numerator and the denominator by the same factor.

Example

$\frac{12}{28} = \frac{3}{7}$ dividing the numerator and the denominator by 4

or $\frac{54}{72} = \frac{27}{36} = \frac{9}{12} = \frac{3}{4}$ dividing top and bottom by 2 then by 3 and then by 3 again.

You must remember that percentages are fractions with denominators of 100 (per cent means per 100). For example, 5% represents $\frac{5}{100}$, 75% represents $\frac{75}{100}$.

You can convert percentages to decimals by dividing by 100, so 5% = $\frac{5}{100}$ = 0.05, and 75% = $\frac{75}{100}$ = 0.75.

To change a fraction into a percentage first change it into a decimal and then multiply by 100.

Example

$\frac{3}{8}$ as a percentage is 3 ÷ 8 = 0.375, 0.375 × 100 = 37.5%

To find the percentage of a quantity change the % into a decimal and then multiply the result by the quantity.

Example

either find 30% of 50 = 0.3 × 50 = 15

or find 10% of 50 = $\frac{1}{10}$ × 50 = 5 so 30% = 3 × 10% = 3 × 5 = 15

You need to be able to calculate percentages in problems such as 'what is 14 marks out of 25 as a percentage?':

Example

either 14 out of 25 as a fraction is $\frac{14}{25}$ = $\frac{14}{25}$ × 100 = 14 × 4 = 56%

or use equivalent fractions: $\frac{14}{25}$ = $\frac{56}{100}$ (multiplying numerator and denominator by 4 to get a denominator of 100)

= 56%

Example

Percentages are useful for comparisons:

In a test Richard got 40 right out of 80, Sarah got 45% and Paul managed to get $\frac{5}{8}$ correct. Who did best and who did worst in the test?

Richard got 50%; Paul got $\frac{5}{8}$ × $\frac{100}{1}$ = 62.5%

So Sarah did the worst and Paul did the best.

Here are some common fractions, decimals and percentages.

You should learn these.

1%	=	$\frac{1}{100}$	=	0.01	(divide by 100)
5%	=	$\frac{1}{20}$	=	0.05	(divide by 20)
10%	=	$\frac{1}{10}$	=	0.1	(divide by 10)
$12\frac{1}{2}$%	=	$\frac{1}{8}$	=	0.125	(divide by 8)
20%	=	$\frac{1}{5}$	=	0.2	(divide by 5)
25%	=	$\frac{1}{4}$	=	0.25	(divide by 4)
50%	=	$\frac{1}{2}$	=	0.5	(divide by 2)
75%	=	$\frac{3}{4}$	=	0.75	(divide by 4, multiply by 3)

Questions

1. Calculate these totals without using a calculator:

(a) 1.8 + 2.0 + 0.5   (b) 0.4 + 0.04 + 4   (c) 2.1 + 0.09 + 7 + 0.9

(d) 2.8 + 3.2 − 0.6   (e) 0.04 + 1.04 + 0.4   (f) 2.01 + 0.09 + 7 + 0.09

2. Calculate these without using a calculator:

(a) 1.4 × 30   (b) 0.5 × 0.7   (c) 0.4 × 5

3. Write these percentages as fractions in their simplest form:

(a) 2%   (b) 25%   (c) 85%   (d) 12.5%   (e) 47%

4. Write these fractions as percentages:

(a) $\frac{3}{8}$ (b) $\frac{13}{25}$ (c) $\frac{12}{40}$ (d) $\frac{36}{60}$

5. Work these out:

(a) 25% of £40   (b) 75% of £20   (c) 12% of 50   (d) 20% of 45

6. Simplify these fractions, writing them in their lowest terms:

(a) $\frac{24}{36}$ (b) $\frac{18}{30}$ (c) $\frac{30}{75}$ (d) $\frac{75}{100}$

7. Which is largest?

$\frac{90}{150}$, $\frac{39}{60}$, $\frac{61}{100}$

Mean, median, mode and range

The *mean* is the average most people give if asked for an average – the mean is found by adding up all the values in the list and dividing this total by the number of values.

The *median* is the middle value when all the values in the list are put in size order. If there are two 'middle' values the median is the mean of these two.

The *mode* is the most common value.

The *range* is the difference between the highest value and the lowest value.

Example

This example should illustrate the calculations:

The children in Class 6 gained the following marks in a test:

Boys:	45	46	48	60	42	53	47	51
	54	54	49	48	47	53	48	45
Girls:	45	47	47	55	46	53	54	63
	48	50	46	51	48	48		

Work out the mean, median, mode and range for the boys and girls and compare the distributions of the marks.

The calculation for the boys:

Mean: $\dfrac{42+45+45+46+47+47+48+48+48+49+51+53+53+54+54+60}{16}$

= 49 (to the nearest whole number)

Median: there are 16 values, so the median is midway between the 8th and 9th values

$= \dfrac{48 + 48}{2} = 48.$

The *mode* is 48.

The *range* is $60 - 42 = 18$.

Question

8. Now work out the values for the girls. Then compare the distributions.

2 | Mental arithmetic

Notes

The mathematics required in this part of the test should usually be straightforward. The content and skills likely to be tested are listed in the Introduction (see page 1). Look back at this to remind yourself.

> **Key point**
> When you are taking the test, listen for, and jot down, numbers that may give short cuts or ease the calculations, such as those that allow doubling and halving. For example, multiply by 100 then divide by 2 if you need to multiply by 50, or multiply by 100 and divide by 4 if you are multiplying by 25. To calculate percentages, first find 10% by dividing by 10 then double for 20% or divide by 2 for 5% and so on. Look back at the hints in the Introduction.

Remember:

- Calculators are not allowed.
- **Questions will be read out twice. When answering the questions in this section, ask someone to read each question out to you and then, without a pause, read out the question again.**
- **There should then be a pause to allow you to record the answer before the next question is read out. The pause should be 8 seconds long.**

Questions

1. As part of a two and a quarter hour tennis training session pupils received specialist coaching for one hour and twenty minutes. How many minutes of the training session remained?

2. A test has forty questions, each worth one mark. The pass mark was seventy per cent. How many questions had to be answered correctly to pass the test?

3. Dining tables seat six children. How many tables are needed to seat one hundred children?

4. Three classes of twenty-eight pupils took the end of Key Stage Two mathematics test. Sixty-three pupils gained a level five result. What percentage is this?

5. A coach holds fifty-two passengers. How many coaches will be needed for a school party of four hundred and fifty people?

6. Eight kilometres is about five miles. About how many kilometres is thirty miles?

7. The journey from school to a sports centre took thirty-five minutes each way. The pupils spent two hours at the sports centre. They left school at oh-nine-thirty. At what time did they return?

8. It is possible to seat forty people in a row across the hall. How many rows are needed to seat four hundred and thirty-two people?

9. Pupils spent twenty-five hours in lessons each week. Four hours per week were allocated to science. What percentage of the lesson time per week was spent on the other subjects?

10. A teacher wants to show a twenty-five-minute video. Tidying up the room and setting homework she estimates will take ten minutes. The lesson will finish at eleven forty-five. What is the latest time she can set the video to start?

11. In a test eighty per cent of the pupils in class A achieved level four and above. In class B twenty-two out of twenty-five pupils reached the same standard. What was the difference between the two classes in the percentage of pupils reaching level four and above?

12. Two hundred pupils correctly completed a sponsored spell of fifty words. Each pupil was sponsored at five pence per word. How much money did the pupils raise in total?

13. A pupil scores forty-two marks out of a possible seventy in a class test. What percentage score is this?

14. There are one hundred and twenty pupils in a year group. Each has to take home two notices. Paper costs three pence per copy. How much will the notices cost?

15. What is seven and a half per cent as a decimal?

16. In a class of thirty-five pupils, four out of seven are boys. How many girls are there in the class?

17. In a school there are five classes of twenty-five pupils and five classes of twenty-eight pupils. How many pupils are there in the school?

18. A school has four hours and twenty-five minutes class contact time per day. What is the weekly contact time?

19. In part one of an examination a pupil scored eighteen marks out of a possible twenty-five marks. In part two he scored sixteen marks out of twenty-five. What was his final score for the examination? Give your answer as a percentage.

20. A teacher needs to interview forty pupils for their Record of Achievement. Each pupil is allocated eight minutes. What is the minimum number of half-hour lessons needed to carry out all of the interviews? Interviews must not overlap lessons.

21. A teacher wants to record a film on a three-hour video tape which starts at eleven fifty-five p.m. and ends at one forty-five a.m. the following day. How much time will there be left on the tape?

22. A test has thirty questions, each worth one mark. If the pass mark is sixty per cent, what is the minimum number of questions that must be answered correctly in order to pass the test?

23. A space two point five metres by two point five metres is to be used for a flower bed. What is this area in square metres?

24. In a class of thirty pupils sixty per cent of the pupils are girls. How many boys are there in the class?

25. In a GCSE examination forty-five percent of the national entry of twenty thousand pupils gained a grade C or better. How many pupils was this?

26. Twenty per cent of the pupils in a school with three hundred and fifteen pupils have free school meals. How many pupils is this?

27. Pupils spent twenty-four hours in lessons each week. Six hours per week were spent on design and technology and art lessons. What percentage of lesson time per week was spent on other subjects?

28. A pupil scores fourteen out of a possible twenty-five in a test. What is this as a percentage?

29. Three fifths of a class of thirty-five pupils are boys. How many are girls?

30. A school's end of key stage mathematics test results for a class of twenty-five pupils showed that nineteen pupils achieved level five or above. What percentage was this?

31. What is twelve and one half per cent as a decimal?

32. What is four point zero five six multiplied by one hundred?

33. A school's end of Key Stage two mathematics test results for a class of thirty pupils showed that twenty-four pupils achieved level four and three achieved level 5. What percentage of pupils achieved level 3 or below?

34. A bus journey starts at eight fifty-five. It lasts for forty minutes. At what time does it finish?

35. Twenty per cent of the pupils in Year Ten play hockey. Twenty-five per cent play basketball. The rest play football. There are two hundred pupils in Year Ten. How many play football?

36. A school day finished at fifteen thirty. There were two afternoon lessons of fifty minutes each, with a break of fifteen minutes between the lessons. At what time did the first afternoon lesson begin?

37. A school's end of Key Stage three mathematics test results for a class of thirty-two pupils showed that twenty pupils achieved level six and above. What percentage was this?

38. A teacher travels from school to a training course. After the course is over she returns to school. The distance to the training venue is twenty-four miles and expenses are paid at a rate of forty pence per mile. How much will she receive?

39. A teacher plans to show a twenty-minute video to a group of pupils. The video will be followed by a discussion and then the pupils will take a fifteen-minute test. The lesson will last for fifty minutes. How long can the discussion last?

40. As part of an induction assessment visit an adviser will watch a teacher teach two consecutive lessons of forty minutes each and then spend forty-five minutes talking to the teacher. If the first lesson starts at oh-nine-fifteen, what time will the visit end?

3 | Using and applying general arithmetic

Notes

Many of the questions in the Skills Test will require you to be able to interpret charts, tables and graphs. These are usually straightforward but do make sure that you read the questions carefully and read the tables or graphs carefully so that you will be able to identify the correct information. These questions are so varied that it is difficult to give examples for all of them – practice makes perfect, though.

There are some questions for which you may wish to revise the mathematics:

Fractions and percentages

See the brief notes in Chapter 1 for the essential knowledge. If you have to calculate percentage increases (or decreases), the simplest method is: find the actual difference, divide by the original amount and then multiply by 100 to convert this fraction to a percentage.

Example

Last year 30 pupils gained a level 3 in the national assessment tests. This year 44 gained a level 3. Calculate the percentage increase.

Actual increase = 14.

Percentage increase = $\frac{14}{30} \times 100 = 46.667\%$

Rounding

Clearly this answer, 46.667%, is too accurate. It would be better written as 46.7% (written to 1 decimal place) or as 47% (to the nearest whole number). You need to be able to round answers to a given number of decimal places or to the nearest whole number (depending on what the question is demanding). The simple rule is that if the first digit that you wish to remove is 5 or more, then you add 1 to the last remaining digit in the answer. If the first digit is less than 5 then the digits are just removed.

Examples:
46.3	= 46 to the nearest whole number
0.345	= 0.35 to 2 decimal places
34.3478	= 34.348 to 3 decimal places
34.3478	= 34.35 to 2 decimal places
34.3478	= 34.3 to 1 decimal place
34.3478	= 34 to the nearest whole number.

Ratio and proportion

These sorts of questions are best illustrated with examples:

Example

(i) Divide £60 between 3 people in the ratio $1:2:3$. The total number of 'parts' is $1 + 2 + 3 = 6$.

Therefore 1 part = £60 ÷ 6 = £10.

Therefore the money is shared as £10; £20; £30.

(ii) Four times as many children in a class have school dinners as do not. If there are 30 children, how many have school dinners?

The ratio is $4:1$ giving $4 + 1 = 5$ parts. Therefore 1 'part' = $30 ÷ 5 = 6$.

Therefore $4 \times 6 = 24$ children have school dinners.

Don't confuse ratio and proportion. Ratio is 'part to part' while proportion is 'part to whole' and is usually given as a fraction. If the question asked 'What proportion of children have school dinners?' the answer would be $\frac{24}{30} = \frac{4}{5}$.

Notes on measures

You need to know and be able to change between the main metric units of measurement. For example:

Length   1 kilometre = 1000 metres
1 metre = 100 centimetres or 1000 millimetres
1 centimetre = 10 millimetres

Mass   1 kilogram = 1000 grams
1 tonne = 1000 kilograms

Capacity   1 litre = 1000 millilitres = 100 centilitres

Notes on algebra

Generally a formula will be given to you, either in words or letters, and you will need to substitute numbers into that formula and arrive at an answer through what will be essentially an arithmetic rather than algebraic process. Remember the rules that tell you the order in which you should work through calculations:

- **Brackets should be evaluated first.**
- **Then work out the multiplications and divisions.**
- **Finally work out the additions and subtractions.**

Thus: (i) $2 \times 3 + 4 = 6 + 4 = 10$ but $2 + 3 \times 4 = 14$ (i.e. $2 + 12$) not 20.

(ii) $\frac{6 + 4}{2} = \frac{10}{2} = 5$ not $3 + 4 = 7$.

(iii) $2(3 + 6) = 2 \times 9 = 18$.

Note that some of the questions that follow have several parts. In the actual test each 'part' would be a separate question, e.g. question 1 could be split into three different questions.

Questions

1. There were 30 pupils in a class. Their results in a test are summarised in the table below.

Mark out of 40	Number of pupils achieving mark
19	2
24	8
27	1
29	5
33	2
34	5
36	7

    What are the mean, mode and range for these results?

    In the test this question could be a 'select and place' question as follows:

    Place the correct values in the summary table below.

    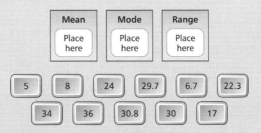

2. A teacher was planning a school trip to Germany. Each pupil was to be allowed €100 spending money. At the time she planned the trip £1 was equivalent to €1.43. When the pupils went to Germany the exchange rate was £1 = €1.38. How much more **English** money did each pupil need to exchange to receive €100?

3. Five times as many pupils in a school obtained level 3 in the Key Stage 2 mathematics test as obtained level 4. If a total of 32 pupils took the test, and just 2 pupils obtained level 5, how many obtained level 3?

4. Teachers in a mathematics department analysed the Key Stage 2 national test results for mathematics from 3 feeder schools.

Level	School A Number of pupils	School B Number of pupils	School C Number of pupils	Totals
2	5	3	4	12
3	6	8	8	22
4	16	18	15	49
5	6	3	8	17
Totals	33	32	35	

    Which school had the greatest percentage of pupils working at level 4 and above?

5. The national percentage of pupils with SEN (including statements) is about 18%. A school of 250 pupils has 35 children on the SEN register. How many children is this below the national average?

6. A small primary school analysed its end of Key Stage 2 results for mathematics for the period 2007–10. The table shows the number of pupils at each level.

Mathematics	Level 5	Level 4	Level 3	Level 2	N
2010	19	15	7		1
2009	14	18	3		
2008	23	21	4		
2007	21	16	5	5	1

Which of the following statements is correct?

(a) The percentage of pupils gaining level 4 or level 5 in 2007 was greater than in 2010.

(b) The percentage of pupils who failed to gain a level 5 was greater in 2009 than in 2008.

(c) The number of pupils gaining level 5 in 2010 was 5 percentage points higher than in 2009.

7. An end of year assessment for a class of 27 Year 10 pupils was planned to take 6 hours. As part of the assessment each pupil required access to a computer for 25% of the time. The school's ICT suite contained 30 computers and could be booked for a number of 40-minute sessions.

How many computer sessions needed to be booked for the class?

8. A pupil achieved a mark of 58 out of 75 for practical work and 65 out of 125 on the written paper. The practical mark was worth 60% of the final mark and the written paper 40% of the final mark. The minimum mark required for each grade is shown below.

Grade	Minimum mark
A*	80%
A	65%
B	55%
C	45%
D	35%

What was the grade achieved by this pupil?

9. A pupil obtained the following marks in three tests.

In which test did the pupil do best?

Test 1	Test 2	Test 3
$\frac{45}{60}$	$\frac{28}{40}$	$\frac{23}{30}$

10. The bar chart below shows the marks for a Year 7 test.

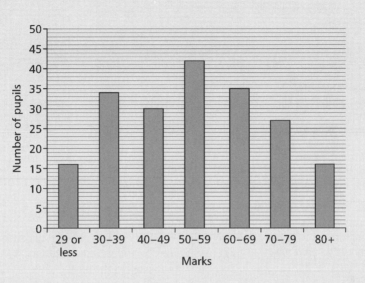

The pass mark for the test was 60 marks. What percentage of pupils passed the test?

11. This table shows the national benchmarks for level 4 and above at Key Stage 2:

Percentile	95th	Upper quartile	60th	40th	Lower quartile	5th
English	97	87	82	73	66	46
Mathematics	98	86	80	72	64	45
Science	100	96	93	87	81	63

Schools' results are placed into categories:

A* Within the top 5%.

A Above the upper quartile and below the 95th percentile.

B Above the 60th percentile and below the upper quartile.

C Between the 40th and the 60th percentiles.

D Below the 40th percentile and above the lower quartile.

E Below the lower quartile and above the 5th percentile.

E* Below the 5th percentile.

What grades would be given for the core subjects in a school whose results were:

English   98%
Mathematics 85%
Science   83%

12. A teacher analysed the number of pupils in a school achieving level 4 and above in the end of Key Stage 2 English tests for 2007–10.

	Year			
	2007	2008	2009	2010
Pupils achieving level 4 and above	70	74	82	84
Pupils in year group	94	98	104	110

In each year the school was set a target of 75% of pupils to achieve a level 4 or above in end of Key Stage English tests.

(a) By how many percentage points did the school exceed its target in 2010? Give your answer to the nearest whole number.

(b) In which year was the target exceeded by the greatest margin?

13. The levels gained in mathematics by the Year 6 pupils in a school in the national attainment tests are shown below. The results are given for Classes 6A and 6B.

Level	Class 6A	Class 6B
N	1	3
2	2	3
3	11	7
4	11	14
5	5	1
6	0	0

(a) What percentage of the year group gained level 4 or above?

(b) Which class had the greater percentage gaining level 4 or above?

14. Four schools had the following proportion of pupils with special educational needs:

School	Proportion
P	$\frac{2}{9}$
Q	0.17
R	57 out of 300
S	18%

Which school had the lowest proportion of pupils with special needs?

(a) School P   (b) School Q   (c) School R   (d) School S

15. This table shows the marks gained by a group of pupils in Year 3 in a mathematics test.

Pupil	Marks	Pupil	Marks
A	23	K	47
B	62	L	38
C	58	M	22
D	35	N	24
E	42	O	81
F	49	P	39
G	76	Q	65
H	80	R	71
I	23	S	73
J	62	T	25

The school will use the results to predict their levels for mathematics at the end of Year 6, and will target those pupils who, it is predicted, will miss level 4 by 1 level.

This is the conversion chart the school uses to change marks to expected levels.

Mark range	20–24	25–51	52–79	80 and over
Expected level	2	3	4	5

How many pupils will be targeted?

16. A school has analysed the results of its students at GCSE and A-level for several years and from these produced a graph which it uses to predict the average A-level points score for a given average points score at GCSE.

Use the graph below to predict the points score at A-level if the GCSE points score were 6.

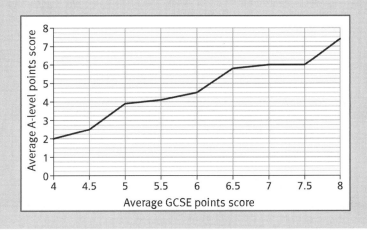

17. A junior school has a weekly lesson time of 23.5 hours. Curriculum analysis gives the following amount of time to the core subjects:

English: 6 hours 30 mins

Mathematics: 5 hours

Science: 1 hour 30 mins

Calculate the percentage of curriculum time given to English. Give your answer to the nearest per cent.

18. A support teacher assessed the reading ages of a group of 10 Year 8 pupils with Special Educational Needs.

Pupil	Actual age		Reading age	
	Years	Months	Years	Months
A	12	07	10	08
B	12	01	11	09
C	12	03	9	07
D	12	03	13	06
E	12	01	10	02
F	12	11	12	00
G	12	06	8	04
H	12	07	10	00
I	12	06	11	08
J	12	02	10	10

What percentage of the 10 pupils had a reading age of at least 1 year 6 months below the actual age?

19. A teacher analysed pupils' performance at the end of Year 5.

Pupils judged to have achieved level 3 and below were targeted for extra support.

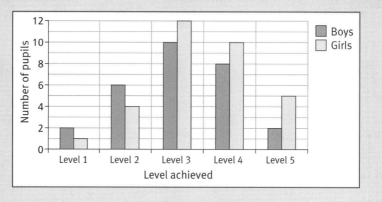

What fraction of the pupils needed extra support?

20. A plastic drinking cup has a capacity of 100ml.

    How many cups could be filled from a 1.5 litre carton of juice?

21. A teacher recorded the number of laps of a rectangular field walked by pupils in Years 5 and 6 in a school's annual walk for charity.

Year group	Number of pupils	Number of laps
5	65	8
6	94	10

    The rectangular field measured 200 metres by 150 metres.

    The teacher calculated the total distance covered.

    Which of the following shows the total distance in kilometres?

    (a) 1022   (b) 1460   (c) 10220   (d) 111.3

22. A primary teacher required each pupil to have a piece of card measuring 20 cm by 45 cm for a lesson. Large sheets of card measured 60 cm by 50 cm. What was the minimum number of large sheets of card required for a class of 28 pupils?

23. For a GCSE German oral examination 28 pupils had individual oral assessments with a language teacher.

    Each individual assessment took 5 minutes. There was a changeover time of 2 minutes between each assessment.

    A day was set aside for the assessments to take place with sessions that ran from 09:00 to 10:10, 10:45 to 11:55 and 13:15 to 15:00.

    At what time did the last pupil finish?
    (a) 14:09   (b) 14:00   (c) 13:43   (d) 13:45

24. A teacher produced this table for pupils in Year 5 showing their predicted levels in English, mathematics and science in the end of Key Stage 2 tests in Year 6.

Pupil	English	Mathematics	Science
A	2	3	3
B	5	5	5
C	5	3	5
D	5	3	4
E	4	3	3
F	4	4	4
G	4	3	5
H	4	4	3

    What proportion of the pupils were predicted to gain level 4 or above in all three subjects in the tests?

    Give your answer as a fraction in its lowest terms.

25. Using the relationship 5 miles = 8 km, convert:

    (a) 120 miles into kilometres.

    (b) 50 km into miles.

    (Give your answers to the nearest whole number in each case.)

26. A ream of photocopier paper is approximately 5 cm thick. What is the approximate thickness of 1 sheet of paper? Give your answers in millimetres.

27. A teacher organised revision classes for pupils who achieved grades C and D in mock examinations and used the following table to assess the number of pupils who might benefit from attending the classes.

Grade	Boys	Girls	Total
A*	3	5	8
A	4	3	7
B	6	8	14
C	7	4	11
D	4	5	9
E	3	3	6
F	2	2	4
G	1	0	1

What percentage of the pupils would benefit from attending the classes?

Give your answer to the nearest whole number.

28. A piece of fabric measuring 32 cm by 15 cm was required for each pupil in a Year 8 design and technology lesson. What was the minimum length of 120 cm wide fabric required for 29 pupils?

29. A school trip is organised from Derby to London – approximately 120 miles. A teacher makes the following assumptions:

    (a) The pupils will need a 30-minute break during the journey.

    (b) The coach will be able to average 40 miles an hour, allowing for roadworks and traffic.

    (c) The coach is due in London at 9 a.m.

    What would be the latest time for the coach to leave Derby?

30. A teacher organised a hike for a group of pupils during a school's activity week. The route was measured on a 1:50 000 scale map and the distances on the map for each stage of the hike were listed on the chart below:

Stage of hike		Distance on map (cm)
1.	Start to Stop A	14.3
2.	Stop A to Stop B	8.7
3.	Stop B to Stop C	9.3

What was the total distance travelled on the hike?

Give your answer to the nearest kilometre.

31. The following table shows the time for 4 children swimming in a relay race:

1st length	John	95.6 seconds
2nd length	Karen	87.3 seconds
3rd length	Julie	91.3 seconds
4th length	Robert	89.4 seconds

What was the total time, in minutes and seconds, that they took?

32. A teacher completed the following expenses claim form after attending a training course:

Travelling

From	To	Miles	Expenses
School and return	Training centre to school	238	place here
Other expenses	Car parking		£7.50
	Evening meal		£10.50
		Total claim	place here

The mileage rates were:

   30p per mile for the first 100 miles

   26p per mile for the remainder.

Complete the claim form by placing the correct values in the expenses column.

£40.88	£65.88	£71.38	£73.88	£83.88	£87.00	£89.40

33. A classroom assistant works from 9:00 a.m. until 12 noon for 4 days per week in a primary school and has a 15 minute break from 10:30 until 10:45. She provides learning support for pupils – each pupil receiving a continuous 20 minutes' session. How many pupils can she support each week?

34. A map has a scale of 1 cm to 6 km. A road on the map is measured as 7.2 cm long. How long is the road in kilometres?

35. A school is expecting 250 parents for a concert. Chairs are to be put out in rows in the hall. Each chair is 45 cm wide and chairs are fixed together in blocks of 8, two blocks making a row. There must be a gangway down the middle of the hall of 0.9 m between the blocks. How much space is needed for each row?

36. Equipment for a school is delivered in boxes 15 cm deep. The boxes are to be stacked in a cupboard which is 1.24 m high. How many layers of boxes will fit into the cupboard?

37. A teacher arranged for four groups of pupils to try out a new interactive program on the classroom computer. He gave a 15-minute introduction and then each group in turn had 10 minutes working at the computer. The changeover time between groups was 2 minutes. How long did the session last?

38. A teacher planned a school trip from Calais to a study centre. The distance from Calais to the centre is 400 km. The coach is expected to travel at an average speed of 50 miles per hour, including time for breaks. The coach is due to leave Calais at 06:20. What time should it arrive at the study centre?

    Use the conversion rate of 1 km = $\frac{5}{8}$ mile.

    Give your answer using the 24 hour clock.

39. A newly qualified teacher in his first year of teaching was given a Year 11 class of 20 pupils. As part of his preparation for a parents' evening he studied a table of end-of-term test results for the class.

Class 11Y – End-of-term test marks (%)					
	Year 10			Year 11	
Name	Test 1	Test 2	Test 3	Test 4	Test 5
A	39	40	41	47	
B	42	44	47	50	
C	42	45	46	50	
D	47	49	48	51	
E	46	48	53	52	
F	50	52	53	57	
G	59	57	58	57	
H	53	55	55	57	
I	55	57	57	61	
J	66	63	65	64	
K	61	60	63	66	
L	56	61	66	68	
M	56	61	66	68	
N	64	66	67	69	
O	51	57	63	69	
P	74	67	69	73	
Q	60	65	71	74	
R	71	73	76	79	
S	75	76	78	81	
T	73	77	82	85	

(i) Indicate all the true statements:

   (a) all pupils showed an improvement in achievement between test 1 and test 4

   (b) the median percentage mark for test 4 was 65%

   (c) the range of percentage marks for test 3 was greater than for test 4.

(ii) The teacher expects the trend of improvement shown by pupil O will continue in the same way for test 5.

Write down the correct value for the improvement for pupil O.

Note: *In the actual test this question would be worded: 'Select and place the correct value for the improvement for pupil O in the test 5 column in the spreadsheet'.*

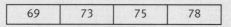

| 69 | 73 | 75 | 78 |

40. Moderators sample the coursework marked by teachers in school. A moderator will select a sample from a school according to the guidelines and rules. One rule that fixes the size of the sample to be selected is:

Size $(s) = 10 + \dfrac{n}{10}$ where n is the number of candidates in a school.

What would be the sample size if there were 150 candidates?

41. A headteacher produced the following table showing Year 11 GCSE results at her school for 2009 and 2010.

2009			5 or more grades A* – C	5 or more grades A* – G	1 or more grades A* – G
GCSE Results					
Number of pupils achieving standard specified	Boys		60	124	126
	Girls		78	108	120
	Total		138	232	246
2010			5 or more grades A* – C	5 or more grades A* – G	1 or more grades A* – G
GCSE Results					
Number of pupils achieving standard specified	Boys		70	130	131
	Girls		75	133	140
	Total		145	263	271

The number of pupils in Year 11 in this school were as follows:

	2009	2010
Boys	126	131
Girls	120	140
Total	246	271

Indicate all the true statements:

(a) in both years all pupils in the school achieved at least one GCSE grade A*– G

(b) the percentage of girls achieving 5 or more GCSE grades A* – G rose by 5 percentage points from 2009 to 2010

(c) a higher percentage of pupils achieved 5 or more GCSE grades A* – C in 2010 than in 2009.

42. As part of a target-setting programme a teacher compared the marks for 10 pupils in each of 2 tests.

Pupils' marks out of 120

Pupil	Test 1	Test 2
A	70	66
B	61	68
C	64	63
D	56	41
E	70	78
F	60	59
G	64	77
H	72	80
I	39	44
J	62	57

Write down the letters for those pupils who scored at least 5 percentage points more in test 2 than test 1.

Note: *In the actual test this question would be expressed as 'Indicate by clicking anywhere on the rows which pupils scored at least 5 percentage points more in test 2 than test 1'.*

43. A pupil achieved the following scores in Tests A, B and C

Test	A	B	C
Actual mark	70	60	7

The pupil's weighted score was calculated using the following formula:

$$\text{Weighted score} = \frac{(A \times 60)}{100} + \frac{(B \times 30)}{80} + C$$

What was the pupil's weighted score?

Give your answer to the nearest whole number.

44. A teacher used a spreadsheet to calculate pupils' marks in a mock GCSE exam made up of two papers. Paper 1 was worth 25% of the total achieved and Paper 2 was worth 75% of the total achieved.

This table shows the first 4 entries in the spreadsheet.

	Paper 1 Mark out of 30	(25%) Weighted mark	Paper 2 Mark out of 120	(75%) Weighted mark	Final weighted mark
Pupil A	24	20	80	50	70
Pupil B	20	16.7	68	42.5	59.2
Pupil C	8	6.7	59	36.9	43.6
Pupil D	20	16.7	74	46.3	63

Pupil E scored 18 on paper 1 and 64 on Paper 2.

What was the final weighted mark for Pupil E?

45. The table below shows the percentage test results for a group of pupils:

Pupil	Test 1	Test 2	Test 3	Test 4	Test 5	Test 6	Test 7	Test 8
A	92	85	87	82	78	26	92	95
B	53	70	72	38	15	27	83	73
C	61	77	69	68	60	30	90	77
D	95	100	93	30	92	30	100	70
E	72	49	47	42	46	82	72	92
F	58	78	38	46	34	58	98	78

Indicate all the true statements:

(a) The greatest range of % marks achieved was in Test 2

(b) Pupil C achieved a mean mark of 66.5%

(c) The median mark for Test 6 was 30.

46. A single mark for a GCSE examination is calculated from three components using the following formula:

Final mark = Component A × 0.6 + Component B × 0.3 + Component C × 0.1.

A candidate obtained the following marks:

Component A   64
Component B   36
Component C   40

What was this candidate's final mark? Give your answer to the nearest whole number.

47. A pupil submitted two GCSE coursework tasks, Task A and Task B.
    Task A carried a weighting of 60% and Task B a weighting of 40%.
    Each task was marked out of 100.
    The pupil scored 80 marks in Task A.

    What would be the minimum mark score required by the pupil in Task B to achieve an overall mark across the two tasks of 60%?

48. A teacher set a pupil a target of achieving a mean score of 70% over four tests.

	Test 1 Out of 30	Test 2 Out of 30	Test 3 Out of 30	Test 4 Out of 30
Score	18	28	14	?

    What was the lowest mark out of 30 that the pupil needed to achieve in Test 4 in order to achieve the target of an overall mean score of 70% that he was set?

49. A teacher calculated the speed in kilometres per hour of a pupil who completed a 6 km cross-country race.

    Use the formula: Distance = speed × time.

    The pupils took 48 minutes.
    What was the pupil's speed in kilometres per hour?

50. A readability test for worksheets, structured examination questions, etc. uses the formula:

    $$\text{Reading level} = 5 + \left\{ 20 - \frac{x}{15} \right\}$$

    Where x = the average number of monosyllabic words per 150 words of writing.
    Calculate the reading level for a paper where $x = 20$. Give your answer correct to 2 decimal places.

51. To help pupils set individual targets a teacher calculated predicted A-level points scores using the following formula:

    $$\text{Predicted A-level points score} = \left(\frac{\text{total GCSE points score}}{\text{number of GCSEs}} \times 3.9 \right) - 17.5$$

    GCSE grades were awarded the following points:

GCSE grade	A*	A	B	C
Points	8	7	6	5

    Calculate the predicted A-level points score for a pupil who at GCSE gained 4 passes at grade C, 4 at grade B, 1 at grade A and 1 at grade A*.

52. A candidate's final mark in a GCSE examination is calculated from two components as follows:

   Final mark = mark in component 1 × 0.6 + mark in component 2 × 0.4

   A candidate needs a mark of 80 or more to be awarded a grade A*. If the mark awarded in component 2 was 70, what would be the least mark needed in component 1 to gain a grade A*?

53. Over a period of years a school has compared performance at GCSE with performance at Key Stage 3 and established rules for the core subjects which they use to predict GCSE grades. In order to do this they converted GCSE grades to points using the following table:

Grade	A*	A	B	C	D	E	F	G
Points	8	7	6	5	4	3	2	1

   For Double Science the school uses the rule:

   GCSE points = Key Stage 3 level − 1

   What would be the expected grade for a candidate who gained a level 5 in science at Key Stage 3?

54. In the annual sports day at a school pupils took part in a running race or in a field event or both. Pupils who took part in both were given an award. In Years 5 and 6 all 72 pupils took part in a running race or in a field event or both.

   $\frac{1}{2}$ took part in a running race and $\frac{3}{4}$ took part in a field event.

   How many pupils were given an award?

55. A school used the ALIS formula relating predicted A-level points scores to mean GCSE points scores for A-level mathematics pupils. The formula used was:

   Predicted A-level points score = (mean GCSE points score × 2.24) − 7.35

   What was the predicted A-level points score for a pupil with a mean GCSE points score of 7.55? Give your answer correct to one decimal place.

56. A teacher compared performance at GCSE with performance at Key Stage 3. The teacher used the following table to convert GCSE grades to points:

Grade	A*	A	B	C	D	E	F	G
Points	8	7	6	5	4	3	2	1

   The formula the teacher used to predict GCSE points was:

   GCSE points = 1.2 × Key Stage 3 level − 2

   What would be the expected grade for a pupil who gained a level 5 at Key Stage 3?

57. Using the same conversion table and the same formula as in the previous question, what was the likely level at Key Stage 3 for a pupil who gained a grade B at GCSE?

58. There were 27 pupils in a Year 8 history class.

One pupil was absent when there was a mid-term test.

The mean score for the group was 56.

On returning to school the pupil who had been absent took the test and scored 86.

What was the revised mean test score?

Give your answer correct to one decimal place.

59. Teachers in a primary school studied the achievement of pupils over a four-year period. End of Key Stage 2 test results of pupils at the school are given in the table below.

	Number of Pupils		
Year	Level 3	Level 4	Level 5
1997	10	60	18
1998	45	60	25
1999	54	48	26
2000	38	58	24

In what year was the ratio of the combined level 3 and level 4 results to the level 5 results exactly 4:1?

60. For a GCSE subject, 20% of the marks were allocated to coursework and 80% to the final examination. The coursework was marked out of 50 and the exam out of 150. A pupil scored 38 for coursework and 112 for the examination.

What was the pupil's final percentage mark for the examination?

Give your answer to the nearest whole number.

4 | Interpreting and using statistical information

Notes

Some terms, concepts and forms of representation which are used in statistics may be unfamiliar. The following notes are intended to give a brief summary of some of the unfamiliar aspects.

Some of the information received by schools, e.g. analyses of pupil performance, uses 'cumulative frequencies' or 'cumulative percentages'. One way to illustrate cumulative frequencies is through an example. The table shows the marks gained in a test by the 60 pupils in a year group:

22	13	33	31	51	24	37	83	39	28
31	64	23	35	9	34	42	26	68	38
63	34	44	77	37	15	38	54	34	22
47	25	48	38	53	52	35	45	32	31
37	43	37	49	24	17	48	29	57	33
30	36	42	36	43	38	39	48	39	59

We could complete a tally chart and a frequency table:

Mark, m	Tally	Frequency
9	1	1
10	0	0
11	0	0
12	0	0
13	1	1
and so on		

But 60 results are a lot to analyse and we could group the results together in intervals. A sensible interval in this case would be a band of 10 marks. This is a bit like putting the results into 'bins':

	13	22 28	33 39,	31, 37		
$0 \leqslant m < 10$	$10 \leqslant m < 20$	$20 \leqslant m < 30$	$30 \leqslant m < 40$	$40 \leqslant m < 50$	etc.

Note that $\leqslant$ means 'less than or equal to' and $<$ means 'less than', so $30 \leqslant m < 40$ means all the marks between 30 and 40 including 30 but excluding 40.

Here are the marks grouped into a frequency table:

Mark, m	Frequency	Cumulative frequency
$0 \leqslant m < 10$	1	1
$10 \leqslant m < 20$	3	4
$20 \leqslant m < 30$	9	13
$30 \leqslant m < 40$	25	38
$40 \leqslant m < 50$	11	49
$50 \leqslant m < 60$	6	55
$60 \leqslant m < 70$	3	58
$70 \leqslant m < 80$	1	59
$80 \leqslant m < 90$	1	60

note how the cumulative frequency is calculated:
$\leftarrow 38 = 1 + 3 + 9 + 25$

The last column 'Cumulative frequency' gives the 'running total' – in this case the number of pupils with less than a certain mark. For example, there are 38 pupils who gained less than 40 marks.

The values for cumulative frequency can be plotted to give a cumulative frequency curve as shown below.

Note that the cumulative frequency values are plotted at the right hand end of each interval, i.e. at 10, 20, 30 and so on.

You can use a cumulative frequency curve to estimate the median mark: the median for any particular assessment is the score or level for which half the relevant pupils achieved a higher result and half achieved a lower result.

There are 60 pupils, so the median mark will be the 30th mark. (Find 30 on the vertical scale and go across the graph until you reach the curve and read off the value on the horizontal scale.) The median mark is about 37 – check that you agree.

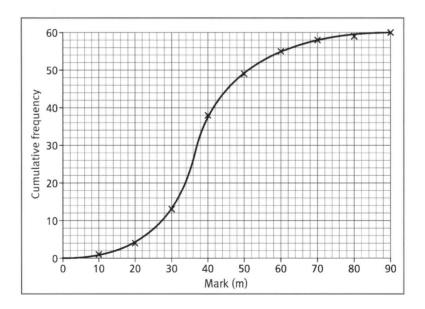

It is also possible to find the quartiles. These are described in the Glossary on page 82.

The lower quartile will be at 25% of 60, that is the 15th value, giving a mark of about 31; the upper quartile is at 75% of 60, thus the 45th value, giving a mark of about 45.

The diagram below should further help to explain these terms. It also helps to introduce the idea of a 'box and whisker' plot.

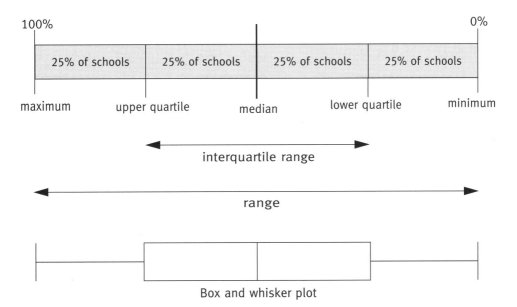

The 'whiskers' indicate the maximum and minimum values, the ends of the 'box' the upper quartile and the lower quartile, and the median is shown by the line drawn across the box.

The 'middle 50%' of the values lie within the box and only the top 25% and the bottom 25% are outside the box. If the ends of the box are close together, then:

- **the upper and lower quartiles are close i.e. the interquartile range (that is the difference between them) is small;**
- **the slope of the cumulative frequency curve (or line) will be steep.**

If the ends of the box are not close, then:

- **the interquartile range is greater;**
- **the data is more 'spread out';**
- **the slope of the curve is less steep.**

You need to interpret 'percentile' correctly: the 95th percentile does not mean the mark that 95% of the pupils scored but that 95% of the pupils gained that mark or lower – it is better perhaps to think that only 5% achieved a higher mark.

The use of percentiles is shown in this table:

Example

Comparing a School's Performance with National Benchmarks, Average NC levels

Percentile	95th	Upper quartile		60th	40th		Lower quartile		5th
English	4.26	4.1	**3.89**	3.89	3.78		3.56		3.36
Mathematics	4.25	3.92		3.85	3.63	**3.63**	3.59		3.24
Science	4.38	4.16		3.91	3.93		3.70	**3.54**	3.49

The figures in bold represent a particular school's average performance.

This table shows, for example, that 5% of pupils nationally gained higher than an average level of 4.26 in the English tests and that 40% of pupils nationally gained an average level of 3.63 or less in mathematics. In other words 60% gained a level higher than 3.63.

The table also shows that the school's performance in English was above average (the 50th percentile) and in line with the 60th percentile, below average in mathematics and well below in science.

Questions

1. A secondary school has compared performance on the Key Stage 2 national tests with performance at GCSE. The comparison is shown on the graph below.

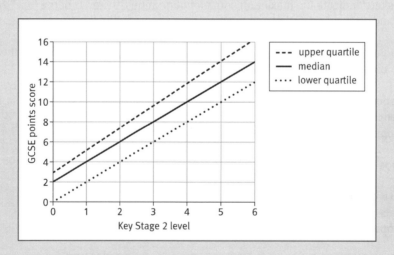

(a) What is the median GCSE points score of those pupils scoring a 2 at Key Stage 2?

(b) A pupil had a Key Stage 2 score of 5 and a GCSE points score of 11. Is it true to say that he was likely to be within the bottom 25% of all pupils?

(c) Is it true that 50% of the pupils who gained level 4 at Key Stage 2 gained GCSE points scores within the range 8 to 12?

2. A teacher produced the following graph to compare performance between an end of Year 9 test and the 2011 GCSE examinations.

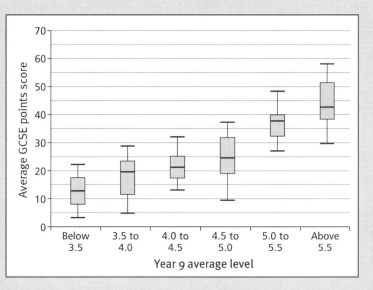

Indicate all the true statements:

1. The median GCSE points score for pupils whose average Year 9 level was 5.0–5.5 was 48.

2. 50% of pupils whose average Year 9 level was 4.0–4.5 achieved average GCSE points scores of between 17 and 25.

3. The range of average GCSE points scores achieved by pupils whose average Year 9 level was above 5.5 was about 28.

3. A German language teacher compared the results of a German oral test and a German written test given to a group of 16 pupils.

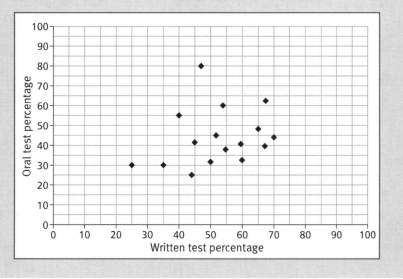

Indicate all the true statements:

1. The range of marks for the oral test is greater than for the written test.

2. $\frac{1}{4}$ of pupils achieved a higher mark on the oral test than on the written test.

3. The two pupils with the lowest marks on the written test also gained the lowest marks on the oral test.

4. A teacher compared the result of an English test taken by all Year 8 pupils

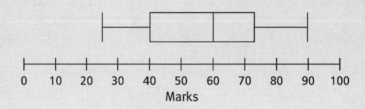

Indicate all the true statements:

1. $\frac{1}{4}$ of all the pupils scored more than 70 marks.

2. $\frac{1}{2}$ of all pupils scored less than 60 marks.

3. The range of marks was 65.

5. In 2010 a survey was made of the nightly TV viewing habits of 10-year-old children in town A and town B. The findings are shown in the pie charts below:

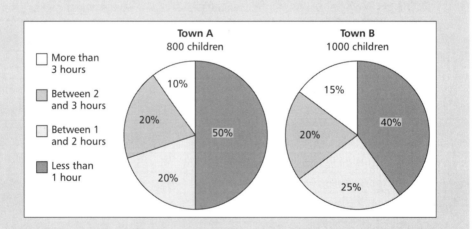

Use these pie charts to identify which of the following statements is true:

1. More children in town A watched TV for less than 1 hour than in town B.

2. More children in town B watched for between 2 and 3 hours than in town A.

3. 100 children watched more than 3 hours in town A.

6. A teacher kept a box and whisker diagram to profile the progress of her class in practice tests. There are 16 pupils in the class.

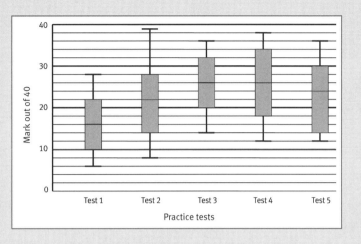

1. In which test did 12 students achieve 20 or more marks?

2. Indicate all the true statements:

   (a) the highest mark was achieved in test 2
   (b) the median mark increased with each test
   (c) the range of marks in test 3 and test 4 was the same.

7. At the beginning of Year 11 pupils at a school took an internal test which was used to predict GCSE grades in mathematics. From the results the predicted grades were plotted on a cumulative frequency graph.

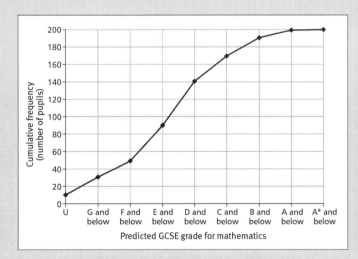

Indicate the true statement:

1. 30% of the pupils were predicted to achieve grade C.

2. 85% of the pupils were predicted to achieve grade C.

3. 15% of the pupils were predicted to achieve grade C.

8. This bar chart shows the amount of pocket money children in Year 7 receive:

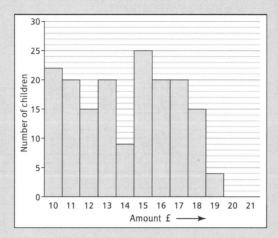

    (a) How many children were surveyed?
    (b) What is the modal amount of pocket money received?

9. The mean height of 20 girls in Year 7 is 1.51 m. Another girl who is 1.6 m joins the class. Calculate the new mean height.

10. A Year 7 teacher was given information from feeder primary schools about pupils in the tutor group.

    The two box plots below show the reading scores for two feeder schools A and B.

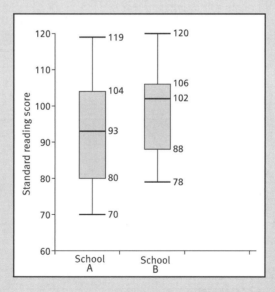

    A standard reading score of 100 shows that a pupil's reading score was exactly on the national average for pupils of the same age. Standard scores of more than 100 show above average reading scores and below 100 show below average reading scores for pupils of the same age.

Indicate all the true statements:

1. The difference in the median scores for the two feeder schools was 9.

2. The interquartile range of the scores for school B was 18.

3. The range of scores was 9 less for school B than for school A.

11. Use the box plots and the information from question 10 to indicate the true statements:

1. 50% of the pupils in school A had a reading score of 93 or more.

2. 25% of pupils in school B scored 88 or less.

3. The interquartile range for the two schools was the same.

12. The following bar chart shows the number of students on NVQ courses in 2009 and 2010.

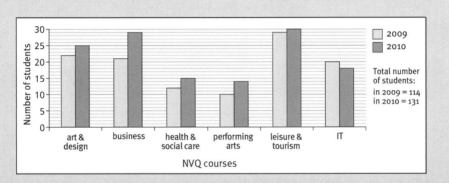

By how many percentage points had the number of students on the NVQ leisure and tourism course changed between 2009 and 2010 when compared with the total number of NVQ pupils for each year? Give your answer to one decimal place.

13. A teacher compared the results of pupils in a school's end of Key Stage 2 mathematics tests.

Results in mathematics	Mean points scored per pupil		
	2008	2009	2010
Boys	22.5	23.2	24.5
Girls	24.3	24.8	25.6

Difference between boys' and girls' mean mathematics points scores			Mean difference for three-year period 2008–2010
2008	2009	2010	
1.8		1.1	

Write down:

(a) the difference between the boys' and girls' mean mathematics points scores for 2009

(b) the mean difference for the three year period 2008–2010.

Note: *In the actual test this question could be worded: 'Place the correct values in the shaded boxes for:*

- *the difference between the boys' and girls' mean mathematics points scores for 2009; and*

- *the mean difference for the three-year period 2008–2010.*

14. This box and whisker diagram shows the GCSE results in 4 subjects for a school in 2011.

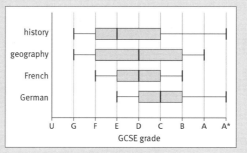

Indicate all the true statements:

1. 50% of the pupils who took history gained grades F to C.
2. French had the lowest median grade.
3. 50% of the pupils who took German gained grades C to A*.

15. The marks of ten students in the two papers of a German examination were plotted on this scattergraph:

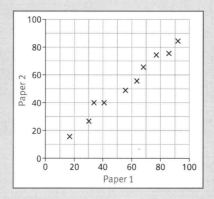

A student scored 53 marks on Paper 1 but missed Paper 2. What would you estimate her mark to be on Paper 2?

16. The straight line shows the mean levels scored in Year 9 assessments for a school plotted against the total GCSE points scores. The points A, B, C, D show the achievement of four pupils.

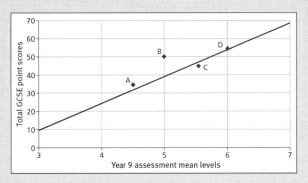

Indicate all the true statements:

1. Pupil D achieved as well as might have been predicted at GCSE.

2. Pupil C achieved a higher level in the Year 9 assessments than Pupil B but scored fewer points at GCSE than Pupil B.

3. At GCSE Pupil B achieved better than might have been predicted but Pupil A achieved less well than might have been predicted.

17. A teacher compared pupil performance in reading in the Key Stage 2 national tests:

| | | | Percentage of Pupils | |
Level achieved	Sex	Year	School	National
3 and above	All	1999	81.8	79.5
3 and above	All	2000	82.3	79.0
3 and above	All	2001	83.4	81.1
3 and above	All	2002	83.6	81.1
3 and above	Boys	2003	78.4	73.9
3 and above	Boys	2004	77.6	74.2
3 and above	Boys	2005	79.0	76.7
3 and above	Boys	2006	79.0	76.4
3 and above	Girls	2007	85.3	84.3
3 and above	Girls	2008	87.1	84.0
3 and above	Girls	2009	88.2	85.7
3 and above	Girls	2010	88.4	85.8

Indicate all the true statements:

1. The performance for the school is consistently above the national average.

2. Girls outperform boys in the reading tests in the school.

3. The percentage of boys achieving level 3 and above increased annually in the school.

18. 20 pupils in a class took Test A at the beginning of a term and Test B at the end of the term.

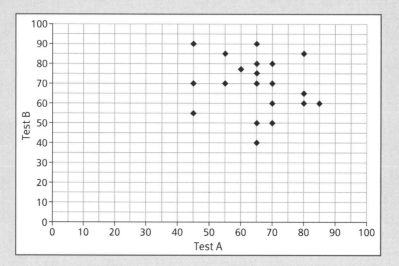

Indicate all the true statements:

1. The range of marks was wider for Test A than for Test B.

2. The lowest mark in Test A was lower than the lowest mark in Test B.

3. 40% of the pupils scored the same mark or lower in Test B than in Test A.

4. More pupils scored over 60% in test A than in test B.

19. A teacher analysed the reading test standardised scores of a group of pupils:

Pupil	Gender	Age 8+ test standardised score	Age 10+ test standardised score
A	F	100	108
B	M	78	89
C	M	88	92
D	M	110	102
E	F	102	110
F	F	88	84
G	M	119	128
H	F	80	84

Indicate all the true statements:

1. All the girls improved their standardised scores between the Age 8+ and the Age 10+ tests.

2. The greatest improvement between the Age 8+ and the Age 10+ tests was achieved by a boy.

3. $\frac{1}{4}$ of all the pupils had lower standardised scores in the Age 10+ tests than in the Age 8+ tests.

20. The graph shows the predicted achievements of pupils in English at the end of Year 9, based on the results of tests taken at the beginning of Year 9.

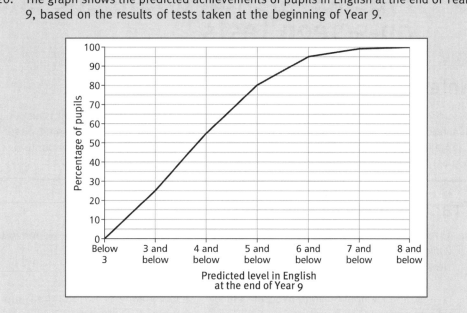

What percentage of pupils was predicted to achieve level 4 and above?

5 | Practice mental arithmetic and on-screen tests

Notes

Before you attempt the tests, re-read the information about the tests in the introduction to this book (see page 1), the hints and advice on pages 2-3, and the opening page of Chapter 2. If you want to time your work, allow yourself 48 minutes to complete both tests. You will find the answers on pages 73–6.

Practice mental arithmetic test

1. Three quarters of a year group of two hundred and forty pupils took part in a sponsored walk for charity.
    How many pupils did not take part?

2. In a GCSE examination forty-five per cent of a school's entry of two hundred pupils gained a grade C or better.
    How many pupils was this?

3. The attendance rate in a school of twelve hundred pupils improves from ninety five per cent to ninety seven per cent in consecutive weeks.
    How many more children were present in the second week?

4. Eight kilometres is about five miles.
    On an activity holiday pupils will cycle between two hostels forty kilometres apart. About how many miles is this?

5. Pupils spent twenty-five hours in lessons each week. Four hours a week were allocated to mathematics.
    What percentage of lesson time per week was spent on other subjects?

6. As part of a practical science workshop some teachers will watch a demonstration lesson of 70 minutes which will be followed by a discussion for 30 minutes. If the demonstration started at oh-nine-thirty, what time will the workshop end?

7. A teacher travels from school to a training course. After the course is over she returns to school. The distance to the training venue is twenty-four miles and expenses are paid at a rate of forty pence per mile. How much will she receive?

8. Pupils spent twenty four hours in lessons each week. Six hours per week were spent on design and technology and art lessons. What percentage of the week is this?

9. A school's end of Key Stage two mathematics test results for a class of thirty pupils showed that twenty five pupils achieved level four and two achieved level 5. What percentage of pupils achieved level 3 or below?

10. A school minibus travels two hundred miles. Fuel costs twenty-one pence per mile. What was the cost of the journey?

11. A teacher completed a claim form for the number of miles travelled to and from five training sessions. The journey one way was fourteen and a half miles. What was the total number of miles claimed?

12. Two thirds of a class of 27 pupils are judged to be on target for a grade C result in their GCSE mathematics examination. How many pupils are judged to get a grade D result or worse?

Practice on-screen test

1. In preparation for target setting a teacher at a comprehensive school produced a scatter graph showing comparative GCSE results for 2010 and 2011 for 10 schools, labelled A to J, in the local authority.

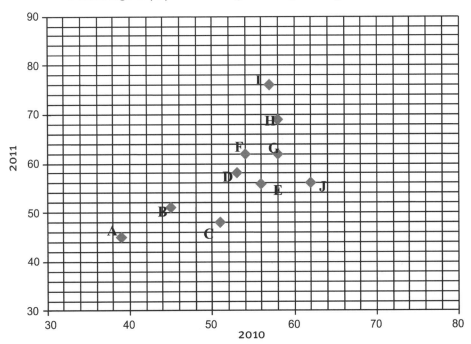

Percentage of pupils achieving at least 5 A*– C grades at GCSE

Write down the letters of the two schools whose percentage of pupils achieving at least 5 A*–C grades at GCSE decreased by more than 2% between 2010 and 2011.

2. A careers teacher produced the following pie charts showing the other subjects taken by sixth form students who also chose to study A level Mathematics.

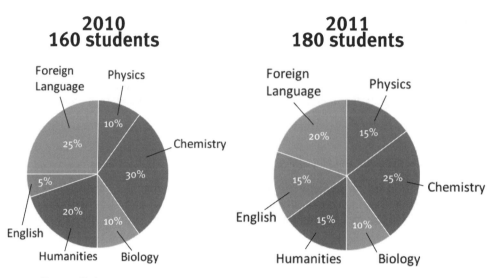

Indicate all the true statements:

1. More pupils chose physics in 2010 than in 2011.
2. The same number of pupils chose chemistry in each year.
3. 4 more pupils chose English in 2010 than in 2011.

3. Pupils in a primary school raise money for charity by collecting tokens from packets of cereal bars. They receive £2.50 for every complete set of 250 tokens collected, with a bonus of £10 for 2000 tokens collected.

How much do they raise if they collect 5140 tokens?

4. A teacher analysed some pupils' results in their mock GCSE examinations in Year 11. She produced the following table showing their predicted grades for English and mathematics

| Subject | | Number of Pupils achieving each grade | | | |
		Grade D	Grade C	Grade B	Grades A/A*
English	Boys	14	34	32	10
	Girls	13	37	35	15
Mathematics	Boys	28	30	16	9
	Girls	10	45	23	20

Indicate all the true statements:
1. The percentage of boys predicted to gain a Grade C in their GCSE mathematics examination is greater than the percentage of boys predicted to gain a grade C in their English examination.
2. A higher percentage of girls than boys are predicted to gain a grade C or grade B in mathematics.
3. The percentage of boys predicted to gain a grade D in mathematics is double the percentage of boys predicted to gain a grade D in English.

5. A local authority gave a primary school a target of 75% of pupils to achieve level 4 and above in end of Key Stage 2 tests in mathematics.

    The proportion of pupils actually achieving each level in end of Key Stage 2 mathematics tests is given in the table below.

Level	2	3	4	5
Proportion	0.1	0.1	0.6	0.2

    By how many percentage points did the school exceed its target?

6. A teacher produced the following chart to show performance of Year 11 pupils in GCSE mathematics in 2011.

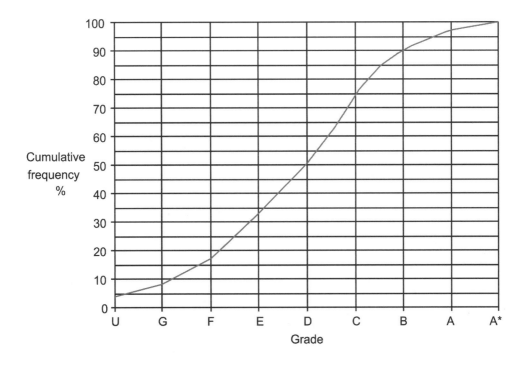

    There were 240 pupils in the year group.
    What was the number of pupils who achieved a grade B and above in mathematics?

    1. 10
    2. 60
    3. 25

7. The performance in a writing task of pupils in different year groups was determined as part of a research project.
A teacher produced the following diagram to show the levels children reached in writing in the different year groups.

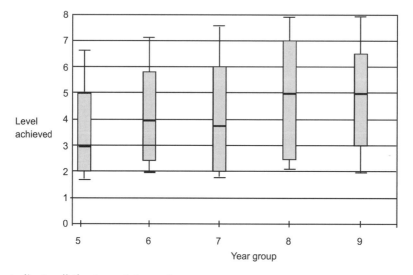

Indicate all the true statements:

1. The median writing level decreased between year 8 and Year 9.
2. In Year 9, 50% of pupils achieved a writing level of 5 or more.
3. In Year 7, 25% of pupils achieved a writing level of 2 or less.

8. The scatter graph below shows the end of year test results in mathematics and science for a group of year 7 pupils.

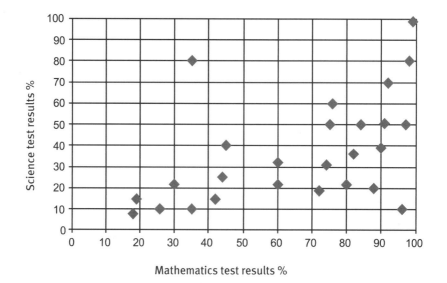

How many pupils achieved more than 60% in at least one of the tests?

9. A head of sixth form uses the following formula to predict A level achievement in points:

$$\text{Predicted A level performance} = \frac{20 \times \text{total GCSE points score}}{\text{number of GCSEs taken}} - 90$$

Student	Total GCSE points score	Number of GCSEs taken	Predicted A level points score, rounded to the nearest whole number
A	28	5	
B	56	8	50
C	45	7	39

21	22	23
92	93	94

What would be the predicted points score for student A?

Write (Select and Place) the correct predicted A level performance for Pupil A in the table.

10. For a trip to Germany pupils have a spending money allowance of £150.
A pupil returns with 30 Euros. How much of his allowance did he spend?
Use the fact that £1 is equivalent to 1.2 Euros.

11. In a GCSE modular examination the uniform mark grade boundaries for each unit are shown below:

Unit	A*	A	B	C	D	E	F	G	U
Foundation Tier 1				60	50	40	30	20	0
Higher Tier 1	90	80	70	60	50	45			
Foundation Tier 2				60	50	40	30	20	0
Higher Tier 2	90	80	70	60	50	45			
Foundation Tier 3				120	100	80	60	40	0
Higher Tier 3	180	160	140	120	100	90			

A pupil's uniform mark for each component will be added together to give the total uniform mark.

The following table gives the minimum total uniform mark required for each overall grade.

	A*	A	B	C	D	E	F	G	U
Total uniform mark	360	320	280	240	200	160	120	80	0

A pupil's mark for the higher tier unit 1 was 65, his mark for the higher tier unit 2 was 75.

What mark does he need to gain for the higher tier 3 unit in order to gain a pass at GCSE grade B?

12. A primary school headteacher prepared a chart showing the percentage of pupils who achieved level 4 and above in mathematics in the end of Key Stage 2 tests.

Percentage of pupils achieving level 4 and above in Key Stage 2 mathematics.

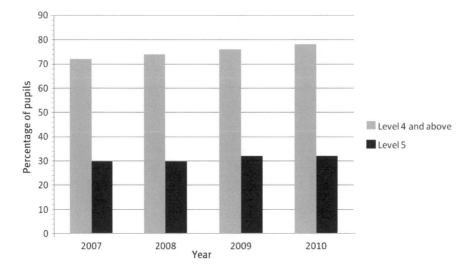

Indicate all the true statements:

1.   In each year less than half the pupils achieved below level 5.
2.   In 2010 22% of pupils achieved below level 4.
3.   If the trend of achievement continues the percentage of pupils achieving level 4 and above will be 80% in 2012.

13. A sixth form history teacher plans to take a group of 90 students and 6 adults to the site of an archaeological excavation for a guided tour and then to the associated museum.

The admission charges are as follows:

	Adults	Students
Excavation tour	£5.60	£3.20
Museum	£4.60	£2.80
Combined admission charge	£9.00	£4.50

One adult is admitted free for every complete group of 20 students.

The group visit both the excavation and the museum.

How much is saved if they buy the combined tickets?

14. A teacher prepares a bar chart to compare the percentage of pupils achieving GCSE grades A* to C in mathematics with other GCSE subjects.

    140 pupils sat the GCSE mathematics examination and 90 achieved a GCSE grade A*–C.

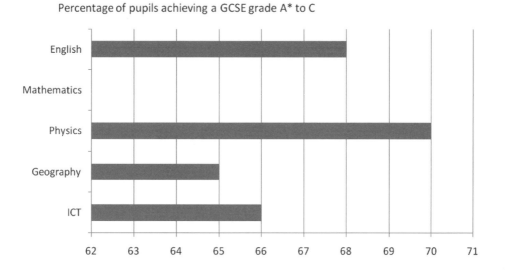

Percentage of pupils achieving a GCSE grade A* to C

Which one of the following bars ought to be placed on the bar chart above to represent the mathematics results?

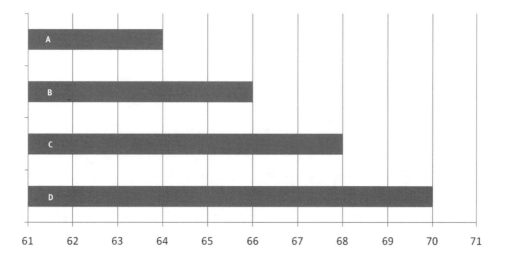

15. This table shows the GCSE grades in design and technology achieved by a school's Year 11 pupils for the period 2008 to 2011.

Grade	A*	A	B	C	D	E	F/G	Total number of students
2008	1	2	2	11	5	6	1	28
2009	6	5	9	13	1	0	0	34
2010	4	7	9	16	4	0	0	40
2011	5	6	8	10	4	2	1	36

Which of the following statements is correct?

1   2009 had the highest percentage of grades A and A*.
2   2008 had the lowest percentage of grade C.
3   Less than ¼ of pupils gained a grade B in any year.

16. A school report includes grades for each pupil's attainment in each subject. Grade A is awarded for an average test mark of 72% and above.

Pupil	Mark
A	31
B	34
C	12
D	17
E	29
F	19
G	24
H	30
I	32
J	28
K	25
L	33

Pupil	Mark
M	7
N	15
O	31
P	23
Q	26
R	28
S	29
T	33
U	29
V	34
W	30
X	28

The table above shows the results for an English test for a group of 24 pupils. If the test was marked out of 36 how many pupils achieved a grade A?

6 | Answers and key points

Chapter 1 Key knowledge

1. (a) 4.3   (b) 4.44   (c) 10.09   (d) 5.4   (e) 1.48   (f) 9.19

> **Key point**
> Remember to line up the decimal points.

2. (a) 42   (b) 0.35   (c) 2

3. (a) $\frac{2}{100} = \frac{1}{50}$ (b) $\frac{25}{100} = \frac{1}{4}$ (c) $\frac{85}{100} = \frac{17}{20}$ (d) $\frac{12.5}{100} = \frac{1}{8}$ (e) $\frac{47}{100}$

> **Key point**
> Cancel down by dividing both numerator and denominator by the same factor.

4. (a) $\frac{3}{8} \times \frac{100\%}{1} = 3 \times 12.5\% = 37.5\%$ (b) $\frac{13}{25} \times \frac{100\%}{1} = 52\%$

   (c) $\frac{12}{40} \times \frac{100\%}{1} = \frac{3}{10} \times \frac{100\%}{1} = 30\%$ (d) $\frac{36}{60} \times \frac{100\%}{1} = \frac{6}{10} \times \frac{100\%}{1} = 60\%$

> **Key point**
> There are different ways of simplifying – which way is best depends on the numbers in the question, but if the denominator is a multiple of 10, it is probably easier to try to cancel down to get 10, as in (c) and (d), unless you are using a calculator in which case change the fraction into a decimal then multiply by 100.

5. (a) $0.25 \times £40 = £10$ (b) $0.75 \times £20 = £15$ (c) $0.12 \times 50 = 6$ (d) $0.2 \times 45 = 9$

> **Key point**
> Change each percentage into a decimal.

6. (a) $\frac{2}{3}$ divided by 12, or by 2, then 2 then 3   (b) $\frac{3}{5}$ divided by 6

   (c) $\frac{2}{5}$ divided by 15, or by 3, then 5   (d) $\frac{3}{4}$ divided by 25

7. $\frac{39}{60}$

> **Key point**
> Change each fraction to a decimal.

8. Girls:

    mean   = 50 (to the nearest whole number)
    median   = 48
    mode   = 48
    range   = 63 − 45 = 18

For both boys and girls the median, mode and range are the same but the mean for the girls is slightly higher so one could deduce that the girls are slightly better than the boys, but the difference is not significant.

Key point
Make sure you put the values in order before finding the median. See the working out for the boys if you have any difficulties.

Chapter 2 Mental arithmetic

1. 55 minutes

Key point
Count on from 1 hour 20 minutes

2. 28

Key point
70% of 40 = 0.7 × 40 or $\frac{7}{10}$ × 40

3. 17

Key point
Note 100 ÷ 6 = 16.666 therefore 17

4. 75%

Key point
The calculation is 3 × 28 = 84, then (63 ÷ 84) × 100. Note: in a mental test the fractions will cancel easily, as here to $\frac{3}{4}$

5. 9

Key point
Note 450 ÷ 52 = 8.65 therefore 9

6. 48

Key point
Calculate as 30 ÷ 5 = 6, then 6 × 8 = 48 or 8 ÷ 5 = 1.6 so 30 × 1.6 = 3 × 16 = 48

7. 12:40

Key point
Count on 1 hr 10 minutes + 2 hours

8. 11

> **Key point**
>
> 10 rows for 400 people, so one more row needed for the remaining 32

9. 84%

> **Key point**
>
> 21 hours remaining = $\frac{21}{25}$ = 84%

10. 11:10 am

> **Key point**
>
> Video + tidying needs 25 + 10 = 35 mins, so 11.45 − 35 = 11.10

11. 8%

> **Key point**
>
> Remember: to convert fractions with denominators of 25 to a percentage, multiply the numerator by 4: $\frac{22}{25}$ = 88%, therefore 8% difference

12. £500

> **Key point**
>
> Simplify quickly by changing into £: 200 × 50 × 5p = £2 × 50 × 5

13. 60%

> **Key point**
>
> Common factor is 7, therefore $\frac{42}{70}$ = $\frac{6}{10}$ = 60%

14. 720p or £7.20

> **Key point**
>
> Work out as 120 × 2 × 3 = 120 × 6

15. 0.075

> **Key point**
>
> Think of $7\frac{1}{2}$% as 7.5% then divide by 100

16. 15

> **Key point**
>
> Work out as $\frac{3}{7}$ × 35

17. 265

> **Key point**
> $(5 \times 25) + (5 \times 28)$

18. 22 hours and 5 minutes

> **Key point**
> $5 \times (4hr\ 25\ mins) = (5 \times 4hr) + (5 \times 25\ min) = 20hr + 125mins = 20hr + 2hr\ 5\ mins$

19. 68%

> **Key point**
> $18 + 16 = 34$ then double

20. 14

> **Key point**
> Think of the context: 3 pupils per half-hour lesson. $\therefore$ $40 \div 3 = 13\ r\ 1$ so 14 half-hour periods

21. 1 hr 10 mins

> **Key point**
> The calculation is $3hr - 1hr\ 50\ mins$

22. 18

> **Key point**
> Find 60% of 30

23. $6.25m^2$

> **Key point**
> The calculation is 2.5×2.5. You should know that 25^2 is 625

24. 12

> **Key point**
> Quicker to find 40% or $0.4 \times 30 = 12$

25. 9000

> **Key point**
> The calculation is $0.45 \times 20\ 000$

26. 63

> **Key point**
> The calculation is $0.2 \times 315 = 63$

27. 75%

> **Key point**
> 18 hours on other subjects so $\frac{18}{24}$ as a percentage

28. 56%

> **Key point**
> With 25 as the denominator you should know that you multiply the numerator by 4

29. 14

> **Key point**
> Find $\frac{1}{5}$ and double to give $\frac{2}{5} \times 35 = 14$

30. 76%

> **Key point**
> $\frac{19}{25}$, multiply 19 by 4 to get the %

31. 0.125

> **Key point**
> Write it as a fraction and divide by 100

32. 405.6

> **Key point**
> Simple to multiply by 100 but be careful!

33. 10%

> **Key point**
> $24 + 3 = 27$ so only 3 didn't get L4 or L5

34. 9:35

> **Key point**
> Treat 40 minutes as 30 minutes + 10 minutes, i.e. 8:55 → 9:25 → 9:35

35. 110

> **Key point**
> 20% + 25% = 45% so 55% play football

36. 13:35

> **Key point**
> 50 + 50 + 15 = 115 so 5 minutes less than 2 hours

37. 62.5%

> **Key point**
> $20 \div 32 \rightarrow$ simplify to $\frac{5}{8}$. You should know that $\frac{1}{8}$ is 12.5% so $\frac{5}{8}$ is 62.5%

38. £19.20

> **Key point**
> The calculation is $2 \times 24 \times 0.4$

39. 15 minutes

> **Key point**
> 20 + 15 = 35 minutes

40. 11:20

> **Key point**
> 40 + 40 + 45 gives 2 hours 5 minutes

Chapter 3 General arithmetic

1. Mean = 29.7; mode = 24; range = 17

> **Key point**
> $$\text{Mean} = \frac{33 \times 2 + 29 \times 5 + 34 \times 5 + 36 \times 7 + 19 \times 2 + 24 \times 8 + 27 \times 1}{30} = \frac{890}{30} = 29.7$$
>
> Mode = most frequent mark, not the number of times it occurs.

2. £2.53

> **Key point**
> 100 euros = $£\frac{100}{1.43}$ = £69.93 and 100 euros = $£\frac{100}{1.38}$ = £72.46

3.   5 gained level 3

> **Key point**
> If 2 gained level 5 then 30 gained either level 3 or level 4. The ratio of level 3 numbers to level 4 numbers is 5 to 1. Divide 30 by (5 + 1), i.e. into 6 equal groups so each group has 5 pupils so 25 pupils gained level 3 and 5 gained level 4.

4.   School A

> **Key point**
> Percentages are: School A = 66.7%; school B = 65.6%; school C = 65.7%

5.   10 below

> **Key point**
> Find 18% of 250 = 0.18 × 250 = 45, so 10 below

6.   false, true, true

> **Key point**
> Looking at actual pupil numbers can be misleading since the totals vary each year. So, in 2007, for statement (a), the fraction is $\frac{37}{48}$ and in 2010 it is $\frac{34}{42}$, but the percentages are 77% and 81% respectively.

7.   3

> **Key point**
> 25% of 6 hours is 1 h 30 min so 2 sessions (80 mins) is too short

8.   A

> **Key point**
> The calculation is $(\frac{58}{75} \times 0.6) + (\frac{65}{125} \times 0.4) = 0.464 + 0.208 = 0.672$ ie 67.2%

9.   3

> **Key point**
> Convert to percentages

10. 39%

> **Key point**
> Total number of pupils = 16 + 34 + 30 + 42 + 35 + 27 + 16 = 200
> Pupils with 60 marks or more = 35 + 27 + 16 = 78

11. English A*   Mathematics B   Science D

Key point

You need to read both question and table carefully.

12. (a) 1%   (b) 2009

Key point

(a) in 2010 $\frac{84}{110}$ = 76% (b) you will have to work out and compare the percentages for the years 2007–10 – you have already worked out the value for 2010.

13. (a) 53.4%   (b) class 6B

Key point

Easy to make mistakes by confusing 'year group' and 'class'.
Total for each class: class A = 30 pupils, class B = 28 pupils
Total for year = 58 pupils

(a) Total pupils gaining level 4 or better = 11 + 5 + 14 + 1 = 31
    As a percentage = $\frac{31}{58}$ × 100 = 53.4%

(b) In class A the percentage = $\frac{16}{30}$ × 100 = 53.3%

    in class B the percentage = $\frac{15}{28}$ × 100 = 53.6%, therefore answer is class B

14. School Q

Key point

Change each figure into decimals:
$\frac{2}{9}$ = 0.222   57 out of 300 = 0.19   18% = 0.18

15. 7

Key point

Look at table 2. To miss level 4 by 1 level thus gaining level 3 means you need to identify pupils scoring marks between 25 and 51.

16. 4.5

Key point

Use a ruler to help – find 6 on the horizontal axis and read off the corresponding value on the vertical axis.

17. 28%

Key point

Because the times both involve half hours, it is simply working out $\frac{6.5}{23.5}$ × 100 = 27.66% = 28% to the nearest per cent.

18. 50%

> **Key point**
> Subtract, remembering 12 months in a year. Pupils A, C, E, G, H fit the criterion.

19. $\frac{35}{60} = \frac{7}{12}$

> **Key point**
> Count pupil numbers carefully – jot down totals.

20. 15

> **Key point**
> $1500 \div 100 = 15$

21. a

> **Key point**
> Number of 'pupil laps' = $(65 \times 8 + 94 \times 10) = 1460$
> Total distance = $1460 \times 700 = 1\,022\,000$ metres = 1022 km

22. 10

> **Key point**
> Use a sketch – 3 small sheets per width of large sheet.

23. 14:09

> **Key point**
> Remember last pupil doesn't need 2 minutes' changeover time.

24. $\frac{1}{4}$

> **Key point**
> The question says 'level 4 or above'. Remember to simplify the fraction.

25. (a) 192 km   (b) 31 miles (actually 31.25)

> **Key point**
> The calculations are: (a) $120 \times \frac{8}{5}$ (b) $50 \times \frac{5}{8}$

26. 0.1 mm

> **Key point**
> Take care with the units – work in millimetres, i.e. $50 \div 500$.

27. 33%

> **Key point**
>
> The fraction is $\frac{20}{60}$ so as a percentage this is 33.3333%, i.e. 33% to the nearest whole number.

28. 128 cm

> **Key point**
>
> You can fit 8 lots of 15 cm across the 120 cm width.

29. 5:30 am

> **Key point**
>
> Remember time = distance ÷ speed. The travel time = 120 ÷ 40 = 3 hr. Add 0.5 hour, therefore total time = 3.5 hours.

30. 16 km

> **Key point**
>
> Total map distance = 32.3 cm = 32.3 × 50 000 cm on the ground = 16.15 km

31. 363.6 seconds = 6 minutes 3.6 seconds

> **Key point**
>
> Add up the time in seconds and decimals of seconds giving 363.6 seconds then convert.

32. £65.88; £83.88

> **Key point**
>
> Remember to work in £ on the mileage rate.

33. 28 pupils

> **Key point**
>
> From 9.00 to 10.30 is 90 minutes → she can see 4 pupils.
>
> From 10.45 to 12.00 is 75 minutes → she can see 3 pupils.
>
> Total for the day = 7 pupils, total over 4 days = 28 pupils.
>
> If the calculations were done using the total figures:
>
> her working week = 4 × 3 hours less 4 × 15 minutes = 11 hours
>
> 11 hours = 660 minutes ÷ 20, giving 33 pupils
>
> This would be incorrect because it ignores the 'structure' of the school morning.

34. 43.2 km

> **Key point**
> 1 cm = 6 km, therefore 7.2 cm = 7.2 × 6 km

35. 8.1 m

> **Key point**
> The fact that 250 parents are coming is redundant information. Working in metres the calculation is 2 × 8 × 0.45 + 0.9

36. 8 layers

> **Key point**
> The calculation, working in centimetres, is 124 ÷ 15 = 8.266, so round down.

37. 61 minutes

> **Key point**
> The calculation is 15 + 4 × 10 + 3 × 2

38. 11:20

> **Key point**
> 400 km = 250 miles

39. (i) (a) false   (b) true   (c) true

    (ii) 75

> **Key point**
> The test scores for test 4 are in order so the median is midway between the 10th and 11th pupils' scores. The range is the difference between the highest and lowest scores.
> In (ii) the scores increase by 6 each time.

40. 25

> **Key point**
> Sample size = 10 + 150 ÷ 10 = 25

41. (a) and (b) are true

> **Key point**
> For statement (a): it is clear all pupils gained a GCSE
> For statement (b): in 2009 the fraction of girls is $\frac{108}{120}$, which gives 90%, and in 2010 it is $\frac{133}{140}$, which gives 95%
>
> For statement (c): check if $\frac{138}{246}$ is greater or less than $\frac{145}{271}$

42. B, E, G, H

> **Key point**
> Find 5% of 120 = 6 so look for marks in test 2 that are 6 marks higher than in test 1.

43. 72

> **Key point**
> Remember to work out brackets first and to round up.

44. 55

> **Key point**
> You can check the method by working through the data for pupil A.
> The calculation for pupil E is $\frac{18}{30} \times 0.25 + \frac{64}{120} \times 0.75 = 0.15 + 0.4 = 0.55 = 55\%$

45. (b) and (c) are true

> **Key point**
> Greatest range is test 5 (92 − 15 = 77)

46. 53

> **Key point**
> The calculation is $64 \times 0.6 + 36 \times 0.3 + 40 \times 0.1 = 53.2$, i.e. 53

47. 30

> **Key point**
> The calculation is $80 \times 0.6 + M \times 0.4 = 60$ so $M \times 0.4 = 60 - 48 = 12$
> Therefore $M = \frac{12}{0.4} = 30$

48. 24

> **Key point**
> Total possible score = 120. 70% of 120 = 84. Therefore marks should add to 84.
> Therefore test 4 mark = 24. OR mean score of 70% = $\frac{21}{30}$. Therefore total marks =
> $4 \times 21 = 84$ so test 4 mark is 24

49. 7.5 km/h

> **Key point**
> Speed = distance ÷ time = 6 ÷ 0.8 (NB work in hours, 48 mins = $\frac{4}{5}$ hr = 0.8 hr)

50. 23.67

> **Key point**
> You must work out $20 \div 15$ first, not 25×20 then divide by 15.
> Therefore reading level $= 5 + (20 - 1.33) = 5 + 18.67 = 23.67$

51. 5.51

> **Key point**
> The total points are given by $(4 \times 5) + (4 \times 6) + (1 \times 7) + (1 \times 8) = 59$
> The calculation is then $(\frac{59}{10} \times 3.9) - 17.5 = 5.51$

52. 87

> **Key point**
> The calculation is $80 = A \times 0.6 + 70 \times 0.4$
> $\qquad\qquad\qquad\quad 80 = A \times 0.6 + 28$
> Therefore $\qquad\quad A \times 0.6 = 52$
> $\qquad\qquad\qquad\quad A = 52 \div 0.6 = 86.67$, i.e. 87

53. Grade D

> **Key point**
> The calculation is: GCSE points $= 5 - 1 = 4$. Therefore Grade $= D$

54. 18

> **Key point**
> $\frac{3}{4} - \frac{1}{2} = \frac{1}{4}$ so $\frac{1}{4}$ receive an award.

55. 9.6

> **Key point**
> Work out brackets first.

56. Grade D

> **Key point**
> The points are given by $1.2 \times 5 - 2 = 4$; therefore, from the table, 4 points $=$ Grade D

57. Level 7

> **Key point**
> From the table a Grade B gives 6 points.
> The calculation is $6 = 1.2 \times L - 2$. Thus $1.2 \times L = 8$ and $L = 6.66$, i.e. level 7

58. 57.1

> **Key point**
>
> Total for 26 pupils = 26 × 56 = 1456. New total = 1456 + 86 = 1542
> New mean = 1542 ÷ 27

59. 2000

> **Key point**
>
> Multiplying each level 5 score by 4 should help identify the year.

60. 75%

> **Key point**
>
> Find 20% of $\frac{38}{50}$ = 15.2% and find 80% of $\frac{112}{150}$ = 59.7%, total = 74.9%

Chapter 4 Interpreting and using statistical information

1. (a) 6   (b) No   (c) Yes

> **Key points**
>
> (a) Find 2 on the KS2 axis and move up the graph until you reach the median line. Read off the value on the GCSE axis.
>
> (b) Find 11 on the GCSE axis and 5 on the KS2 axis. The lines through these values intersect in the space between the lower quartile line and the median line so it is not true – remember that 25% of the pupils are below the lower quartile.
>
> (c) Find 4 on the KS2 axis, 50% lie between 8 (the lower quartile) and 12 (the upper quartile).

2. 2 and 3 are true

> **Key point**
>
> 1. The median is the line inside the 'box'.
> 2. 50% is represented by the box.
> 3. range = highest − lowest = 58 − 30 = 28

3. 1 and 2 are true

> **Key points**
>
> 1. Oral range = 80 − 25 = 55; written range = 70 − 25 = 45.
>
> 2. Imagine a line drawn from (0, 0) to (100, 100). This is the line where scores on both tests were the same. There are 4 points above this line, $\frac{4}{16} = \frac{1}{4}$
>
> 3. Lowest written marks are 25 and 35, both scoring 30 in the oral but one pupil scored 25 in the oral.

4. 2 and 3 are true

> **Key points**
>
> 1. The upper quartile is at 73.
>
> 2. Median is at 60 so true.
>
> 3. Range = 90 − 25 = 65 so true.

5. Only 2 is true

> **Key points**
>
> It is important to realise that, although the pie charts appear to be the same size, they represent different 'quantities' – 800 children and 1000 children.
>
> Statement 1: is not true. 50% of 800 = 40% of 1000
> (0.5 × 800 = 400 and 0.4 × 1000 = 400)
>
> Statement 2: is true. 20% of 1000 is more than 20% of 800
> (0.2 × 1000 = 200, 0.2 × 800 = 160)
>
> Statement 3: is not true. 10% of 800 = 80

6. (i) test 3 (ii) (a) true, (b) false, (c) false

> **Key point**
>
> Remember that each 'part' of a box and whisker plot represents 25%.

7. Only 3 is true

> **Key point**
>
> 140 pupils get D and below, 170 get C and below so 30 get grade C.

8. (a) 170 (b) £15

> **Key point**
>
> (a) Add up the values given by the tops of each bar:
> 22 + 20 + 15 + 20 + 9 + 25 + 20 + 20 + 15 + 4
>
> (b) The modal amount is that received by the most children, i.e. £15 received by 25 children.

9. 1.514 m

> **Key point**
>
> The total height for the 20 girls = 20 × 1.51 = 30.2 m
>
> The new total height = 30.2 + 1.6 = 31.8 m but this is for 21 girls.
>
> The new mean height = 31.8 ÷ 21 = 1.514 m

10. 1 and 2 are true

> **Key point**
>
> 1. Medians are 102 and 93.
>
> 2. Interquartile range for B = 106 − 88
>
> 3. School A range = 119 − 70, school B range = 120 − 78

11. 1 and 2 are true

> **Key point**
>
> 1. 50% lie above the median line.
>
> 2. 25% lie below the lower quartile.
>
> 3. IQ range for A = 24 and for B = 18

12. 2.5%

> **Key point**
>
> Total for 1999 = 114, and for 2000 = 131 − given in question.
>
> 1999 tourism number = 29 and $\frac{29}{114}$ = 25.4%
>
> 2000 number = 30 and $\frac{30}{131}$ = 22.9%

13. 1.6, 1.5

> **Key point**
>
> (a) Simple subtraction. (b) Simple calculation of the mean.

14. 1 and 3 are true

> **Key point**
>
> 1. History 'box' extends from F to C.
> 2. History has a lower median grade.
> 3. German median is at Grade C so 50% gained C to A*.

15. About 45–49 marks

> **Key point**
>
> You need to draw in the line of best fit through the points.

16. 1 and 2 are true

> **Key point**
>
> 1. D lies on the line.
> 2. C is to the right of but below B.
> 3. A is above the line so A also achieved better than might have been predicted.

17. 1 and 2 are true

> **Key point**
> 1. Check the school numbers are always greater than the national figures.
> 2. Check the girls numbers are always greater than the boys.
> 3. Decreased in 1998.

18. 3 and 4 are true

> **Key point**
> 1. Range for A = 85 − 45 = 40, range for B = 90 − 40 = 50
>
> 2. Lowest value for A = 45 and for B = 40
>
> 3. Imagine a line drawn from (0, 0) to (100, 100), 8 pupils on or below the line and $\frac{8}{20} = 40\%$
>
> 4. 14 scored *over* 60% in A and 13 scored *over* 60% in B

19. 2 and 3 are true

> **Key point**
> 1. Pupil F score decreased
>
> 2. Pupil B score increased by 11
>
> 3. 2 pupils, (D, F) had lower scores

20. 75%

> **Key point**
> 25% predicted to gain level 3 and below so 75% predicted to gain more than level 3, i.e. level 4 and above.

Chapter 5 Practice mental arithmetic test answers

Note that units will not have to be entered in the test – the appropriate unit should appear in the answer box.

1. 60

2. 80

3. 24

4. 25 (miles)

5. 84 (%)

6. 11:10

7. (£) 19.20

8. 25 (%)

9. 10 (%)

10. (£) 42

11. 145 (miles)

12. 9

Practice on-screen test answers

1. C and J

    > **Key point**
    > In an actual test this question could be worded as follows: 'Point and click on the letters of the 2 schools ...' You may find it useful to imagine the line where the percentages are equal for each year, i.e. the line joining (30, 30) and (80, 80). On the screen you could use the edge of a piece of paper for the straight line.

2. All the statements are false.

    > **Key point**
    > Note that the pie charts represent different totals.

3. £70

    > **Key point**
    > Note you need to work with multiples of 250 and 2000.

4. 2

    > **Key point**
    > This question refers to percentages. The table gives actual numbers. You will have to calculate the totals for each row.

5. 5%

    > **Key point**
    > The proportion achieving levels 4 and 5 is 0.8. This represents 80%.

6. 2

    > **Key point**
    > 75% gained a grade C or below so 25% gained a grade B, A , A*. 25% of 240 = 60.

7. 2 and 3

    > **Key point**
    > Look at the notes about box plots if you are unsure how to interpret them.

8. 16

> **Key point**
>
> Imagine a vertical line drawn through the 60% maths result – how many pupils are to the right of this? Imagine a line drawn through the 60% science result. How many pupils are above this line? Be sure, however, not to count pupils twice.

9. 22

> **Key point**
>
> Using the simple on-screen calculator you will need to work out the numerator and jot the answer down, then do the division (here divide by 5) and then subtract 90.

10. £125

> **Key point**
>
> 30 euros = £30 ÷ 1.2 = £25. £150 − £25 = £125.

11. 140

> **Key point**
>
> Unit 1 mark was 65, unit 2 mark was 75. Total = 140. For a grade B pass his total uniform mark must be 280 so he needs to score 140 on the higher tier 3 paper.

12. 2

> **Key point**
>
> The trend of achievement suggests that the level 4 figures increase by 2 each year and this would indicate that the 80% figure would be reached in 2011 not 2012.

13. £137.40

> **Key point**
>
> 4 adults will be free. The saving for an adult is £5.60 + £4.60 - £9.00 = £1.20 and for a student is £3.20 + £2.80 − £4.50 = £1.50. Therefore saving = 2 × £1.20 + 90 × £1.50.

14. A

> **Key point**
>
> 90 ÷ 140 = 64.28% so bar A is the best choice.

15. 1

> **Key point**
>
> Note that the table quotes actual numbers not percentages so you will need to work these out. For example, in 2009 11 pupils gained grades A and A*. 11 out of 34 = 32%.

16. 10

> **Key point**
> First find 72% of 36. This is 25.92, i.e. 26 marks. Next count all those whose mark was 26 or more.

Further reading

You can get further help and guidance with mathematical knowledge from other Learning Matters publications such as:

Mooney, C., Hansen, A., Ferrie, L., Fox, S., and Wrathmell, R. (2012) *Primary Mathematics: Knowledge and Understanding* (6th edition). Exeter: Learning Matters.

Details of publications can be found online at *www.uk.sagepub.co.uk/learningmatters*

Glossary

Accuracy The degree of precision given in the question or required in the answer. For example, a length might be measured to the nearest centimetre. A pupil's reading age is usually given to the nearest month, whilst an average (mean) test result might be rounded to one decimal place.

Bar chart A chart where the number associated with each item is shown either as a horizontal or a vertical bar and where the length of the bar is proportional to the number it represents. The length of the bar is used to show the number of times the item occurs, or the value of the item being measured.

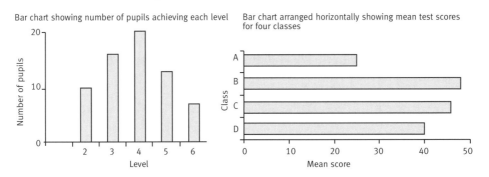

Bar chart showing number of pupils achieving each level

Bar chart arranged horizontally showing mean test scores for four classes

Box and whisker diagram Diagram showing the range and quartile values for a set of data.

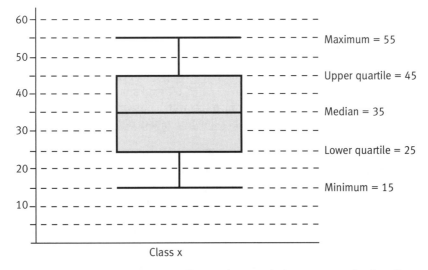

Class x

Cohort A group having a common quality or characteristic. For example, 'pupils studying GCSE German this year have achieved higher grades than last year's cohort' (pupils studying GCSE German last year).

Consistent Following the same pattern or style over time with little change. For example, a pupil achieved marks of 84%, 82%, 88% and 85% in a series of mock GCSE tests; her performance was judged to be consistently at the level needed to obtain GCSE grade A*.

Conversion The process of exchanging one set of units for another. Measurement and currency, for example, can be converted from one unit to another, e.g. centimetres to metres, pounds to euros. Conversion of one unit to the other is usually done by using a rule (e.g. 'multiply by $\frac{5}{8}$ to change kilometres into miles'), a formula (e.g. F = $\frac{9}{5}$C + 32, for converting degrees Celsius to degrees Fahrenheit), or a conversion graph.

Correlation The extent to which two quantities are related. For example, there is a positive correlation between two tests, A and B, if a person with a high mark in test A is likely to have a high mark in test B and a person with a low mark in test A is likely to get a low mark in test B. A scatter graph of the two variables may help when considering whether a correlation exists between the two variables.

Cumulative frequency graph A graph in which the frequency of an event is added to the frequency of those that have preceded it. This type of graph is often used to answer a question such as, 'How many pupils are under nine years of age in a local education authority (LA)?' or 'What percentage of pupils gained at least the pass mark of 65 on a test?'.

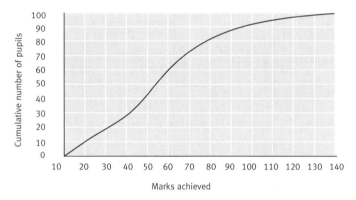

The graph shows the marks pupils achieved. Two pupils scored 10 marks or less, 30 pupils scored 42 marks or less, 60 pupils scored 60 marks or less and 90 pupils scored 95 marks or less. If these were results from a test with a pass mark of 65 marks, then from the graph we can see that 63% of pupils gained 64 marks or less, and so failed the test.

Decimal Numbers based on or counted in a place value system of tens. Normally we talk about decimals when dealing with tenths, hundredths and other decimal fractions less than 1. A decimal point is placed after the units digit in writing a decimal number, eg. 1.25
The number of digits to the right of the decimal point up to and including the final non-zero digit is expressed as the number of decimal places. In the example above there are two digits after the decimal point, and the number is said to have two decimal places, sometimes expressed as 2 dp. Many simple fractions cannot be expressed exactly as a decimal. For example, the fraction $\frac{1}{3}$ as a decimal is 0.3333... which is usually represented as 0.3 recurring. Decimals are usually rounded to a specified degree of accuracy, eg. 0.6778 is 0.68 when rounded to 2 dp. 0.5 is always rounded up, so 0.5 to the nearest whole number is 1.

Distribution The spread of a set of statistical information. For example, the number of absentees on a given day in a school is distributed as follows: Monday – 5, Tuesday – 17, Wednesday – 23, Thursday – 12 and Friday – 3. A distribution can also be displayed graphically.

Formula A relationship between numbers or quantities expressed using a rule or an equation. For example, final mark = (0.6 × mark 1) + (0.4 × mark 2).

Fraction Fractions are used to express parts of a whole, eg. $\frac{3}{4}$. The number below the division line, the denominator, records the number of equal parts into which the number above the division line, the numerator, has been divided.

Frequency The number of times an event or quantity occurs.

Greater than A comparison between two quantities. The symbol $>$ is used to represent 'greater than', eg. $7 > 2$, or $> 5\%$.

Interquartile range The numerical difference between the upper quartile and the lower quartile. The lower quartile of a set of data has one quarter of the data below it and three-quarters above it. The upper quartile has three quarters of the data below it and one quarter above it. The inter-quartile range represents the middle 50% of the data.

Line graphs A graph on which the plotted points are joined by a line. It is a visual representation of two sets of related data. The line may be straight or curved. It is often used to show a trend, such as how a particular value is changing over time.

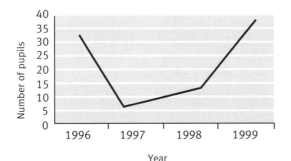

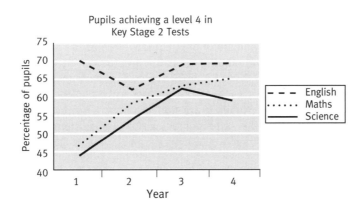

Mean One measure of the 'average' of a set of data. The 'mean' average is usually used when the data involved is fairly evenly spread. For example, the individual costs of four textbooks are £9.95, £8.34, £11.65 and £10.50. The mean cost of a textbook is found by totalling the four amounts, giving £40.44, and then dividing by 4, which gives £10.11. The word average is frequently used in place of the mean, but this can be confusing as both median and mode are also ways of expressing an average.

Median Another measure of the 'average' of a set of data. It is the middle number of a series of numbers or quantities when arranged in order, eg. from smallest to largest. For example, in the following series of number: 2, 4, 5, 7, 8, 15 and 18, the median is 7. When there is an even number of numbers, the median is found by adding the two middle numbers and then halving the total. For example, in the following series of numbers, 12, 15, 23, 30, 31 and 45, the median is (23 + 30) ÷ 2 = 26.5.

Median and quartile lines Quartiles can be found by taking a set of data that has been arranged in increasing order and dividing it into four equal parts. The first quartile is the value of the data at the end of the first quarter. The median quartile is the value of the data at the end of the second quarter. The third quartile is the value of the data at the end of the third quarter.

Quartile lines can be used to show pupils' progression from one key stage to another, when compared with national or local data:

1997/98 Key Stage 3 mathematic test median line (with quartile boundaries)

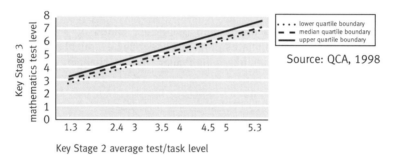

Source: QCA, 1998

Key Stage 2 average test/task level

Mode Another measure of the 'average' of a set of data. It is the most frequently occurring result in any group of data. For example, in the following set of exam results: 30%, 34%, 36% 31%, 31%, 30%, 34%, 33%, 31% and 32%, the mode is 31% because this value appears most frequently in the set of results.

Operations The means of combining two numbers or sets of numbers. For example, addition, subtraction, multiplication and division.

Percentage A fraction with a denominator of 100, but written as the numerator followed by '%', e.g. $\frac{30}{100}$ or 30%. A fraction that is not in hundredths can be converted so that the denominator is 100, e.g. $\frac{650}{1000} = \frac{65}{100} = 65\%$. Percentages can be used to compare different fractional quantities. For example, in class A, 10 pupils out of 25 are studying French; in class B, 12 out of 30 pupils are studying French. However, both $\frac{10}{25}$ and $\frac{12}{30}$ are equivalent to $\frac{4}{10}$, or 40%. The same percentage of pupils, therefore, study French in both these classes.

Percentage points The difference between two values, given as percentages. For example, a school has 80% attendance one year and 83% the next year. There has been an increase of 3 percentage points in attendance.

Percentile The values of a set of data that has been arranged in order and divided into 100 equal parts. For example, a year group took a test and the 60th percentile was at a mark of 71. This means that 60% of the cohort scored 71 marks or less. The 25th percentile is the value of the data such that 25% or one quarter of the data is below it and so is the same as the lower quartile. Similarly, the 75th percentile is the same as the upper quartile and the median is the same as the 50th percentile.

Pie chart A pie chart represents the 360° of a circle and is divided into sectors by straight lines from its centre to its circumference. Each sector angle represents a specific proportion of the whole. Pie charts are used to display the relationship of each type or class of data within a whole set of data in a visual form.

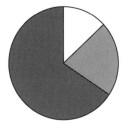

Pie chart showing the distribution of how the total number of pupils in a school take lunch

☐ Packed lunch
▨ Go home
▨ School lunch

Prediction A statement based on analysing statistical information about the likelihood that a particular event will occur. For example, an analysis of a school's examination results shows that the number of pupils achieving A*–C grades in science at a school has increased by 3% per year over the past three years. On the basis of this information the school predicts that the percentage of pupils achieving A*–C grades in science at the school next year will increase by at least 2%.

Proportion A relationship between two values or measures. These two values or measures represent the relationship between some part of a whole and the whole itself. For example, a year group of 100 pupils contains 60 boys and 40 girls, so the proportion of boys in the school is 60 out of 100 or 3 out of 5. This is usually expressed as a fraction, in this case, $\frac{3}{5}$.

Quartile (lower) The value of a set of data at the first quarter, 25%, when all the data has been arranged in ascending order. It is the median value of the lower half of all the values in the data set. For example, the results of a test were: 1, 3, 5, 6, 7, 9, 11, 15, 18, 21, 23 and 25. The median is 10. The values in the lower half are 1, 3, 5, 6, 7 and 9. The lower quartile is 5.5. This means that one quarter of the cohort scored 5.5 or less. The lower quartile is also the 25th percentile.

Quartile (upper) The value of a set of data at the third quarter, 75%, when that data has been arranged in ascending order. It is the median value of the upper half of all the values in the data set. In the lower quartile example, the upper quartile is 19.5, the median value of the upper half of the data set. Three quarters of the marks lie below it. The upper quartile is also the 75th percentile.

Range The difference between the lowest and the highest values in a set of data. For example, for the set of data 12, 15, 23, 30, 31 and 45, the range is the difference between 12 and 45. 12 is subtracted from 45 to give a range of 33.

Ratio A comparison between two numbers or quantities. A ratio is usually expressed in whole numbers. For example, a class consists of 12 boys and 14 girls. The ratio of boys to girls is 12:14. This ratio may be described more simply as 6:7 by dividing both numbers by 2. The ratio of girls to boys is 14:12 or 7:6.

Rounding Expressing a number to a degree of accuracy. This is often done in contexts where absolute accuracy is not required, or not possible. For example, it may be acceptable in a report to give outcomes to the nearest hundred or ten. So the number 674 could be rounded up to 700 to the nearest hundred, or down to 670 to the nearest ten. If a number is half way or more between rounding points, it is conventional to round it up, eg. 55 is rounded up to 60 to the nearest ten and 3.7 is rounded up to 4 to the nearest whole number. If the number is less than half way, it is conventional to round down, eg. 16.43 is rounded down to 16.4 to one decimal place.

Scatter graph A graph on which data relating to two variables is plotted as points, each axis representing one of the variables. The resulting pattern of points indicates how the two variables are related to each other. This type of graph is often used to demonstrate or confirm the presence or absence of a correlation between the two variables, and to judge the strength of that correlation.

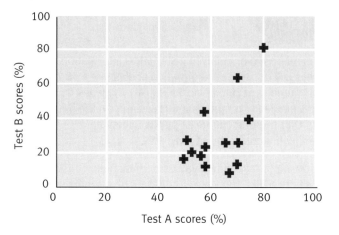

Sector The part or area of a circle which is formed between two radii and the circumference. Each piece of a pie chart is a sector.

Standardised scores Standardised scores are used to enable comparisons on tests between different groups of pupils. Tests are standardised so that the average national standardised scores automatically appear as 100, so it is easy to see whether a pupil is above or below the national average.

Trend The tendency of data to follow a pattern or direction. For example, the trend of the sequence of numbers 4, 7, 11, 13 and 16 is described as 'increasing'.

Value added The relationship between a pupil's previous attainment and their current attainment gives a measure of their progress. Comparing this with the progress made by other pupils gives an impression of the value added by a school. Below is a scatter graph showing progress made by a group of pupils between the end of Key Stage 1 and the end of Key Stage 2:

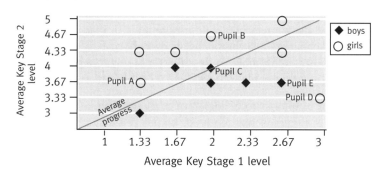

The 'trend line' shows the average performance. Pupils above the line, such as A and B, made better progress than expected; those below the line, such as pupils E and D made less progress than that expected.

How Key Stage 2 relates to Key Stage 1: Median, upper and lower quartile

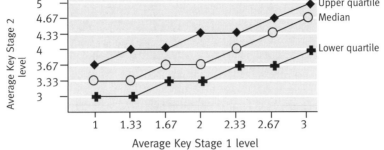

Here the median line shows the national average progress.

Variables The name given to a quantity which can take any one of a given set of values. For example, on a line graph showing distance against time, both distance and time are variables. A variable can also be a symbol which stands for an unknown number, and that can take on different values. For example, the final mark in a test is obtained by a formula using the variables A and B as follows: final mark = (Topic 1 mark × A) + (Topic 2 mark × B).

Weighting A means of attributing relative importance to one or more of a set of results. Each item of data is multiplied by a pre-determined amount to give extra weight to one or more components. For example, marks gained in year 3 of a course may be weighted twice as heavily as those gained in the first two years, in which case those marks would be multiplied by two before finding the total mark for the course.

Whole number A positive integer, eg. 1, 2, 3, 4, 5.

© Teaching Agency